jQuery Mobile Web Development Essentials

Second Edition

Build mobile-optimized websites using the simple, practical, and powerful jQuery-based framework

Raymond Camden

Andy Matthews

[PACKT] open source *
PUBLISHING   community experience distilled

BIRMINGHAM - MUMBAI

jQuery Mobile Web Development Essentials
Second Edition

First published: May 2012

Second Edition: September 2013

Production Reference: 1190913

Published by Packt Publishing Ltd.
Livery Place
35 Livery Street
Birmingham B3 2PB, UK.

ISBN 978-1-78216-789-1

www.packtpub.com

Cover Image by Suresh Mogre (suresh.mogre.99@gmail.com)

Credits

Authors

Raymond Camden

Andy Matthews

Reviewers

Matt Gifford

Eliecer Daza Parra

Olivier Pons

Acquisition Editor

Usha Iyer

Lead Technical Editor

Sweny M. Sukumaran

Technical Editors

Dennis John

Gaurav Thingalaya

Project Coordinator

Apeksha Chitnis

Proofreader

Faye Coulman

Indexer

Mariammal Chettiyar

Production Coordinator

Melwyn D'sa

Cover Work

Melwyn D'sa

About the Authors

Raymond Camden is a Senior Developer Evangelist for Adobe. His work focuses on web standards, mobile development, and ColdFusion. He's a published author, and presents at conferences and user groups on a variety of topics. Raymond can be reached at his blog at www.raymondcamden.com, followed on Twitter (@cfjedimaster), or contacted via e-mail at raymondcamden@gmail.com.

> As always, I dedicate this book to the one person who made this all possible, my wife Jeanne. Thank you for believing in me and being strong when I am not. I will love you always.
>
> I'd like to thank everyone on the jQuery and jQuery Mobile teams for making tools that have changed my life. Without your hard work and dedication, the Web would be less awesome. Thank you Andy for coming on board and helping to make this book better.

Andy Matthews has been working as a web and application developer for over 16 years, with experience in a wide range of industries, and a skillset that includes UI/UX, graphic design, and programming. He is the co-author of the book *jQuery Mobile Web Development Essentials*, *Packt Publishing*, and writes for online publications such as NetTuts and .NET Magazine. He is a frequent speaker at conferences around the country, and he has developed software for the open source community including several of the most popular jQuery Mobile projects on the Web. He blogs at andyMatthews.net, tweets at @commadelimited, and lives in Nashville, TN, with his wife and four children.

> Thanks to my wife and children who tolerate my time spent learning and writing.
>
> Thanks to Packt Publishing for publishing this book. Thanks to the jQuery Mobile team for creating such a great and easy-to-use open source project.

About the Reviewers

Matt Gifford is an RIA developer from Cambridge, England, who specializes in ColdFusion, web application, and mobile development. With over 10 years of industry experience across various sectors. Matt is the owner of a development consultancy firm monkehWorks Ltd (www.monkehworks.com).

He is a regular presenter at national and international conferences, and also contributes articles and tutorials for leading international industry magazines as well as publishes them on his blog at: http://www.mattgifford.co.uk.

As an Adobe Community Professional, Matt is an advocate of community resources and industry-wide knowledge sharing, with a focus on encouraging the next generation of industry professionals.

Matt is the author of *Object-Oriented Programming in ColdFusion* and *PhoneGap Mobile Application Development Cookbook* (both by *Packt Publishing*) as well as numerous open source applications, including the popular monkehTweets Twitter API wrapper.

You can reach Matt on Twitter via @coldfumonkeh or through his blog.

My eternal thanks always go to my constantly supportive family. Big thanks also go to Ray and Andy for inviting me to review their work. It has been a pleasure working with them, as always.

Eliecer Daza Parra has been a web developer since 2005. He has got ample experience in Java, Python, PHP, jQuery, and jQuery Mobile. Elicer has an experience of more than 8 years as a Java developer. He has been a software developer for Information Management and Customer Relationship Management (CMR) for health promoting enterprises (EPS), public transportation, and education companies in the private and public sectors. He has been working as a Python developer since more than 2 years, working with responsive websites.

Among the main areas of his interest are the development of Linux, Python, Android, and Google Services. He has a huge interest in nurturing blog spaces about Linux administration and programming.

My heartfelt appreciation to God, my beloved mother and friend, my family, and July.

Olivier Pons is a developer who's been building websites since 1997. He's a teacher at Ingésup (École supérieure d'ingénierie informatique), the University of Sciences (IUT) of Aix-en-Provence, France where he teaches Linux, Apache HTTP server, PHP, jQuery/jQuery Mobile, advanced website optimization, and advanced VIM techniques. He has already written some technical reviews, including the Packlib book *Ext JS 4 First Look*. In 2011, he left a full-time job as a Delphi and PHP developer to concentrate on his own company, HQF Development (`http://hqf.fr`). He currently runs a number of websites, including `http://www.livrepizzas.fr`, `http://www.papdevis.fr`, and `http://olivierpons.fr`—his own web development blog. He currently works as a consultant, project manager, and senior developer.

www.PacktPub.com

Support files, eBooks, discount offers and more

You might want to visit www.PacktPub.com for support files and downloads related to your book.

Did you know that Packt offers eBook versions of every book published, with PDF and ePub files available? You can upgrade to the eBook version at www.PacktPub.com and as a print book customer, you are entitled to a discount on the eBook copy. Get in touch with us at service@packtpub.com for more details.

At www.PacktPub.com, you can also read a collection of free technical articles, sign up for a range of free newsletters and receive exclusive discounts and offers on Packt books and eBooks.

http://PacktLib.PacktPub.com

Do you need instant solutions to your IT questions? PacktLib is Packt's online digital book library. Here, you can access, read and search across Packt's entire library of books.

Why Subscribe?

- Fully searchable across every book published by Packt
- Copy and paste, print and bookmark content
- On demand and accessible via web browser

Free Access for Packt account holders

If you have an account with Packt at www.PacktPub.com, you can use this to access PacktLib today and view nine entirely free books. Simply use your login credentials for immediate access.

Table of Contents

Preface

Welcome to *jQuery Mobile Web Development Essentials, Second Edition*. Both myself and Andy Matthews have tried our best to create a book that introduces and prepares you for building mobile-friendly websites with jQuery Mobile.

What is jQuery Mobile?

On August 11, 2010, John Resig (creator of jQuery) announced the jQuery Mobile project. While focused on the UI framework, it was also a recognition of jQuery itself as a tool for mobile sites, and that work would be done to the core framework itself to make it work better on devices. Release after release, the jQuery Mobile project evolved into a powerful framework encompassing more platforms, more features, and better performance with every update.

But what do we mean when we refer to a "UI framework"? What does it mean for developers and designers? jQuery Mobile provides a way to turn regular HTML (and CSS) into mobile-friendly sites. As you will see in the book, you can take a regular HTML page, add in the required bits for jQuery Mobile (essentially five lines of HTML), and find your page transformed into a mobile-friendly version instantly.

Unlike other frameworks, jQuery Mobile is focused on HTML. In fact, for a framework tied to jQuery, you can do a heck of a lot of work without writing a single line of JavaScript. It's a powerful, practical way of creating mobile websites that any existing HTML developer can pick up and adapt within a few hours. Compare this to other frameworks, such as Sencha Touch; Sencha Touch is also a powerful framework, but its approach is radically different, using JavaScript to help define and lay out pages. jQuery Mobile is much friendlier to people who are more familiar with HTML as opposed to JavaScript. jQuery Mobile is "touch-friendly", which will make sense to anyone who has used a smart phone, and struggled to click the right spot on a website with tiny text and hard-to-spot links.

It will make sense to anyone who accidentally clicked a Reset button instead of Submit. jQuery Mobile will enhance your content to help solve these issues. Regular buttons become large, fat, and easy to hit. Links can be turned into list-based navigation systems. Content can be split into virtual pages with smooth transitions. You will be surprised just how much jQuery Mobile will do for you without writing much code at all.

jQuery Mobile has some very big sponsors. They include Nokia, Blackberry, Adobe, and other large companies. These companies have invested money, hardware, and developer resources to help ensure the success of the project.

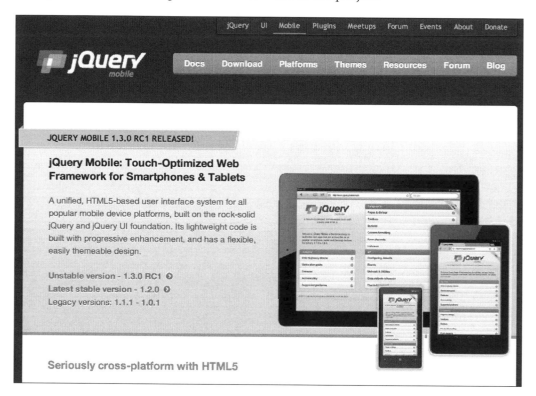

What's the cost?

Ah, the million dollar question! Luckily this one is easy to answer: nothing. jQuery Mobile, like jQuery itself, is completely free to use for any purpose. Not only that, it's completely open source. Don't like how something works? You can change it. Want something not supported by the framework? You can add it. To be fair, digging deep into the code base is probably something most folks will not be comfortable doing. However, the fact that you can if you need to, and the fact that other people can, leads to a product that will be open to development by the community at large.

What do you need to know?

Finally, along with not paying a dime to download and work with jQuery Mobile, the best thing is that you probably already have all the skills necessary to work with the framework. As you will see in the chapters of the book, jQuery Mobile is a HTML-based framework. If you know HTML, even just simple HTML, you can use jQuery Mobile framework. Knowledge of CSS and JavaScript is a plus, but not entirely required (While jQuery Mobile uses a lot of CSS and JavaScript behind the scenes, you don't actually have to write any of this yourself!).

What about native apps?

jQuery Mobile does not create native applications. You'll see later in the book how you can combine jQuery Mobile with "wrapper" technologies such as PhoneGap to create native apps, but in general, jQuery Mobile is for building websites. The question on whether to develop a website or a mobile app is not something this book can answer. You need to look at your own business needs and see what will satisfy them. Because we are not building mobile apps themselves, we do not have to worry about setting up any accounts with Google or Apple or paying any fees for the marketplace. Any user with a mobile device that includes a browser will be able to view your mobile-optimized sites.

Again, if you want to develop true mobile apps with jQuery Mobile, it's definitely an option.

Help!

While we'd like to think that this book will cover every single possible topic you would need for all your jQuery Mobile needs, there will most likely be things we can't cover. If you need help, there are a couple of places you can try.

Firstly, the jQuery Mobile docs (`http://jquerymobile.com/demos/`) cover syntax, features, and development in general, much like this book. While the material may cover some of the same ground, if you find something confusing here, try the official docs. Sometimes a second explanation can really help.

Secondly, the jQuery Mobile forum (`http://forum.jquery.com/jquery-mobile`) is an open-ended discussion list for jQuery Mobile topics. This is the perfect place to ask questions. Also, it's a good place to learn about problems other people are having. You may even be able to help them. One of the best ways to learn a new topic is by helping others.

Examples

Do you want to see jQuery Mobile in action? There's a site for that. JQM Gallery (`http://www.jqmgallery.com/`) is a collection of user-submitted sites built using jQuery Mobile. Not surprisingly, it too uses jQuery Mobile that makes it yet another way to sample jQuery Mobile.

What this book covers

Chapter 1, Preparing Your First jQuery Mobile Project, walks you through your first jQuery Mobile project. It details what must be added to your project's directory and source code.

Chapter 2, Working with jQuery Mobile Pages, continues the work in the previous chapter and introduces the concept of jQuery Mobile pages.

Chapter 3, Enhancing Pages with Headers, Footers, and Toolbars, explains how to enhance your pages with nicely formatted headers and footers.

Chapter 4, Working with Lists, describes how to create jQuery Mobile listviews. These are mobile-optimized lists that are especially great for navigation.

Chapter 5, Getting Practical – Building a Simple Hotel Mobile Site, walks you through creating your first "real" (albeit simple) jQuery Mobile application.

Chapter 6, Working with Forms and jQuery Mobile, walks you through the process of using jQuery Mobile-optimized forms. Layout and special form features are covered in detail.

Chapter 7, Creating Modal Dialogs and Widgets, walks you through special jQuery Mobile user interface items for creating grid-based layouts, dialogs, and collapsible content areas.

Chapter 8, Moving Further with the Notekeeper Mobile Application, walks you through the process of creating another website, an HTML5-enhanced note taking application.

Chapter 9, jQuery Mobile Configuration, Utilities, and JavaScript Methods, describes the various JavaScript-based utilities your code may require.

Chapter 10, Working with Events, details the events thrown by various jQuery Mobile-related features, such as pages loading and unloading.

Chapter 11, Enhancing jQuery Mobile, demonstrates how to change the default appearance of your jQuery Mobile sites by selecting and creating unique themes.

Chapter 12, Creating Native Applications, takes what you've learned previously and shows how to use the open source PhoneGap project to create real native applications.

Chapter 13, Becoming an Expert – Building an RSS Reader Application, expands upon the previous chapter by creating an application that lets you add and read RSS feeds on mobile devices.

What you need for this book

Nothing! Technically, you need a computer and a browser, but jQuery Mobile is built with HTML, CSS, and JavaScript. No IDE (Integrated Development Environment) or special tool will be required to work with the framework. If you've got any editor on your system (and all operating systems include a free editor of some sort), you can develop with jQuery Mobile.

There are good IDEs out there that can help you be more productive. Adobe Dreamweaver CC, for example, includes native support for jQuery Mobile with code assist and device previews.

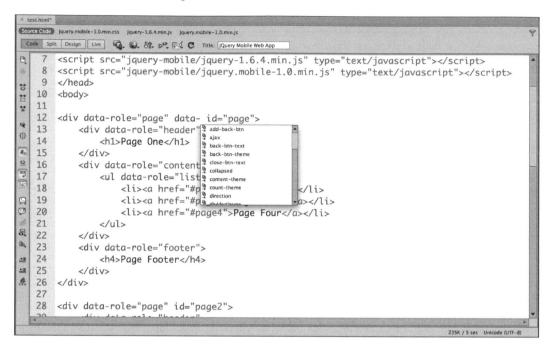

At the end of the day, you can develop with jQuery Mobile for free. It's zero cost for you to download, develop, and publish jQuery Mobile sites.

Who this book is for

This book is for anyone looking to embrace mobile development and expand their skillsets beyond the desktop.

Conventions

In this book, you will find a number of styles of text that distinguish between different kinds of information. Here are some examples of these styles, and an explanation of their meaning.

Code words in text are shown as follows: "Notice the new `data-title` tag added to the `div` tag."

A block of code is set as follows:

```html
<html>
<head>
<title>First Mobile Example</title>
</head>
<body>
```

New terms and **important words** are shown in bold. Words that you see on the screen, in menus or dialog boxes, for example, appear in the text like this: "Imagine our **Megacorp** page. It's got three pages, but the **Products** page is a separate HTML file."

 Warnings or important notes appear in a box like this.

 Tips and tricks appear like this.

Reader feedback

Feedback from our readers is always welcome. Let us know what you think about this book—what you liked or may have disliked. Reader feedback is important for us to develop titles that you can really get the most out of.

To send us general feedback, simply send an e-mail to feedback@packtpub.com, and mention the book title in the subject line of your message.

If there is a topic that you have expertise in and you are interested in either writing or contributing to a book, see our author guide on www.packtpub.com/authors.

Customer support

Now that you are the proud owner of a Packt book, we have a number of things to help you to get the most from your purchase.

Downloading the example code

This book contains many code samples. You are not expected to type them in. You should *not* type them all in. Rather, you should download them from the public GitHub repository set up for the book: `https://github.com/cfjedimaster/jQuery-Mobile-Book`. The GitHub repository will be updated as typos and other mistakes are found in the book. Therefore it is possible that the code may not exactly match the text in the book.

If you are not familiar with Git, then simply click on the **Downloads** tab and then either **Download as zip** or **Download as tar.gz** to quickly get an archived collection of all the files.

You should extract these files onto a local web server. If you do not have one installed, we recommend installing Apache (`http://httpd.apache.org/`). Apache works on all platforms, is free, and is typically easy to install. Once extracted, you can edit these files, view them in your browser, or copy them as a starting point for your own projects.

You can download the example code files for all Packt books you have purchased from your account at `http://www.packtpub.com`. If you purchased this book elsewhere, you can visit `http://www.packtpub.com/support` and register to have the files e-mailed directly to you.

Errata

Although we have taken every care to ensure the accuracy of our content, mistakes do happen. If you find a mistake in one of our books—maybe a mistake in the text or the code—we would be grateful if you would report this to us. By doing so, you can save other readers from frustration and help us improve subsequent versions of this book. If you find any errata, please report them by visiting `http://www.packtpub.com/support`, selecting your book, clicking on the **errata submission form** link, and entering the details of your errata. Once your errata are verified, your submission will be accepted and the errata will be uploaded to our website, or added to any list of existing errata, under the Errata section of that title.

Piracy

Piracy of copyright material on the Internet is an ongoing problem across all media. At Packt, we take the protection of our copyright and licenses very seriously. If you come across any illegal copies of our works, in any form, on the Internet, please provide us with the location address or website name immediately so that we can pursue a remedy.

Please contact us at copyright@packtpub.com with a link to the suspected pirated material.

We appreciate your help in protecting our authors, and our ability to bring you valuable content.

Questions

You can contact us at questions@packtpub.com if you are having a problem with any aspect of the book, and we will do our best to address it.

1
Preparing Your First jQuery Mobile Project

You know what jQuery Mobile is, the history of it as well as its features and goals. Now we're actually going to build our first jQuery Mobile website (well, web page) and see how easy it is to use.

In this chapter we will perform the following steps:

- Create a simple HTML page
- Add jQuery Mobile to the page
- Make use of custom data attributes (`data-*`)
- Update the HTML to make use of the data attributes that jQuery Mobile recognizes

Important preliminary points

You can find all the source code for this chapter in the `c1` folder of the ZIP file you downloaded from GitHub. If you wish to type everything out by hand, we recommend you use similar filenames.

Building an HTML page

Let's begin with a simple web page that is not mobile optimized. To be clear, we aren't saying it can't be viewed on a mobile device. Not at all! But it may not be usable on a mobile device. It may be hard to read (text too small). It may be too wide. It may use forms that don't work well on a touch screen. We don't know what kinds of problems we will have at all until we start testing. (And we've all tested our websites on mobile devices to see how well they work, right?)

Let's have a look at the following code snippet:

```
<h1>Welcome</h1>
<p>
   Welcome to our first mobile web site. It's going to be the
      best site you've ever seen. Once we get some content. And a
      business plan. But the hard part is done!
</p>
<p>
   <i>Copyright Megacorp&copy; 2013</i>
</p>
</body>
</html>
```

As we said, there is nothing too complex, right? Let's take a quick look at this in the browser:

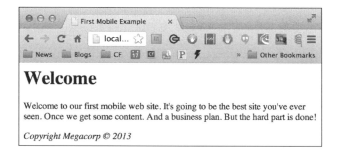

Not so bad, right? But let's take a look at the same page in a mobile simulator:

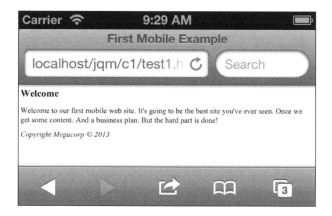

Wow, that's pretty tiny. You've probably seen web pages like this before on your mobile device. You can, of course, typically use pinch and zoom or double-click actions to increase the size of the text. But it would be preferable to have the page render immediately in a mobile-friendly view. This is where jQuery Mobile comes in.

Getting jQuery Mobile

In the preface we talked about how jQuery Mobile is just a set of files. That isn't said to minimize the amount of work done to create those files, or how powerful they are, but to emphasize that using jQuery Mobile means you don't have to install any special tools or server. You can download the files and simply include them in your page. And if that's too much work, you have an even simpler solution. jQuery Mobile's files are hosted on a **Content Delivery Network** (**CDN**). This is a resource hosted by them and guaranteed (as much as anything like this can be) to be online and available. Multiple sites are already using these CDN hosted files. That means when your users hit your site they may already have the resources in their cache. For this book, we will be making use of the CDN hosted files, but just for this first example we'll download and extract the files we need. I recommend doing this anyway for those times when you're on an airplane and wanting to whip up a quick mobile site.

To grab the files, visit `http://jquerymobile.com/download`. There are a few options here but you want the ZIP file option. Go ahead and download that ZIP file and extract it. (The ZIP file you downloaded earlier from GitHub has a copy already.) The following screenshot demonstrates what you should see after extracting the files from the ZIP file:

At the time this book was written, jQuery Mobile was preparing for the release of Version 1.4. Obviously, by the time you read this book a later version may have been released. The file names you see listed in the previous screenshot are version specific, so keep in mind they may look a bit different for you.

Notice the ZIP file contains a CSS and JavaScript file for jQuery Mobile, as well as a minified version of both. You will typically want to use the minified version in your production apps and the regular version while developing. The `images` folder has five images used by the CSS when generating mobile optimized pages. You will also see demos for the framework as well as theme and structure files (You won't need to use those for this book). So, to be clear, the entire framework and all the features we will be talking about over the rest of the book will consist of a framework of 6 files. Of course, you also need to include the jQuery library. You can download that separately at www.jquery.com. At the time this book was written, the recommended version was 1.9.1.

Customized downloads

As a final option for downloading jQuery Mobile, you can also use a customized Download Builder tool at http://jquerymobile.com/download-builder. Currently in Alpha (that is, not certified to be bug-free!), the web-based tool lets you download a jQuery Mobile build minus features your website doesn't need. This creates smaller files which reduces the total amount of time your application needs to display to the end user.

Implementing jQuery Mobile

Ok, we've got the bits, but how do we use them? Adding jQuery Mobile support to a site requires the following three steps at a minimum:

1. First, add the HTML5 DOCTYPE to the page: `<!DOCTYPE html>`. This is used to help inform the browser about the type of content it will be dealing with.

2. Add a viewport `metatag`: `<metaname="viewport"content="width=device-width,initial-scale="1">`. This helps set better defaults for pages when viewed on a mobile device.

3. Finally, the CSS, JavaScript library, and jQuery itself need to be included into the file.

Let's look at a modified version of our previous HTML file that adds all of the above:

```
code 1-2: test2.html
<!DOCTYPE html>
<html>
  <head>
    <title>First Mobile Example</title>
    <meta name="viewport" content="width=device-width, initial-
      scale=1">
    <link rel="stylesheet"href="jquery.mobile-1.3.2.min.css" />
    <script type="text/javascript"
      src="http://code.jquery.com/jquery-1.9.1.min.js"></script>
    <script type="text/javascript"src="jquery.mobile-
      1.3.2.min.js"></script>
  </head>
  <body>
    <h1>Welcome</h1>
    <p>
      Welcome to our first mobile web site. It's going to be the best
        site you've ever seen. Once we get some content. And a
business
        plan. But the hard part is done!
    </p>
    <p>
      <i>Copyright Megacorp&copy; 2013</i>
    </p>
  </body>
</html>
```

For the most part, this version is the exact same as Code 1-1, except for the addition of the DOCTYPE, the CSS link, and our two JavaScript libraries. Notice we point to the hosted version of the jQuery library. It's perfectly fine to mix local JavaScript files and remote ones. If you wanted to ensure you could work offline, you can simply download the jQuery library as well.

So while nothing changed in the code between the body tags, there is going to be a radically different view now in the browser. The following screenshot shows how the iOS mobile browser renders the page now:

Right away, you see a couple of differences. The biggest difference is the relative size of the text. Notice how much bigger it is and easier to read. As we said, the user could have zoomed in on the previous version, but many mobile users aren't aware of this technique. This page loads up immediately in a manner that is much more usable on a mobile device.

Working with data attributes

As we saw in the previous example, just adding in jQuery Mobile goes a long way to updating our page for mobile support. But there's a lot more involved to really prepare our pages for mobile devices. As we work with jQuery Mobile over the course of the book, we're going to use various data attributes to mark up our pages in a way that jQuery Mobile understands. But what are data attributes?

HTML5 introduced the concept of data attributes as a way to add ad-hoc values to the **DOM (Document Object Model)**. As an example, this is a perfectly valid HTML:

```
<div id="mainDiv" data-ray="moo">Some content</div>
```

In the previous HTML, the data-ray attribute is completely made-up. However, because our attribute begins with data-, it is also completely legal. So what happens when you view this in your browser? Nothing! The point of these data attributes is to integrate with other code, like JavaScript, that does whatever it wants with them. So for example, you could write JavaScript that finds every item in the DOM with the data-ray attribute, and change the background color to whatever was specified in the value.

This is where jQuery Mobile comes in, making extensive use of data attributes, both for markup (to create widgets) and behavior (to control what happens when links are clicked). Let's look at one of the main uses of data attributes within jQuery Mobile—defining pages, headers, content, and footers:

```
code 1-3: test3.html
<!DOCTYPE html>
<html>
  <head>
    <title>First Mobile Example</title>
    <meta name="viewport" content="width=device-width, initial-
      scale=1">
    <link rel="stylesheet"href="jquery.mobile-1.3.2.min.css" />
    <script type="text/javascript"src="http://code.jquery
      .com/jquery-1.9.1.min.js"></script>
    <script type="text/javascript"src="jquery.
      mobile-1.3.2.min.js"></script>
  </head>
  <body>
    <div data-role="page">
      <div data-role="header"><h1>Welcome</h1></div>
      <div data-role="content">
        <p>
          Welcome to our first mobile web site. It's going to be the
            best site you've ever seen. Once we get some content. And
            a business plan. But the hard part is done!
        </p>
      </div>
      <div data-role="footer">
        <h4>Copyright Megacorp&copy; 2013</h4>
      </div>
    </div>
  </body>
</html>
```

Compare the previous code snippet to `code 1-2`, and you can see that the main difference was the addition of the `div` blocks. One `div` block defines the page. Notice it wraps all of the content inside the `body` tags. Inside the `body` tag, there are three separate `div` blocks. One has a role of `header`, another a role of `content`, and the final one is marked as `footer`.

All the blocks use `data-role`, which should give you a clue that we're defining a role for each of the blocks. As we stated previously, these data attributes mean nothing to the browser itself. But let's look what at what jQuery Mobile does when it encounters these tags:

Notice right away that both the header and footer now have a black background applied to them. This makes them stand out even more from the rest of the content. Speaking of the content, the page text now has a bit of space between it and the sides. All of this was automatic once the `div` tags with the recognized `data-roles` were applied. This is a theme you're going to see repeated again and again as we go through this book. A vast majority of the work you'll be doing will involve the use of data attributes.

Summary

In this chapter, we talked a bit about how web pages may not always render well in a mobile browser. We talked about how the simple use of jQuery Mobile can go a long way to improving the mobile experience for a website. Specifically, we discussed how you can download jQuery Mobile and add it to an existing HTML page, what data attributes mean in terms of HTML, and how jQuery Mobile makes use of data attributes to enhance your pages. In the next chapter, we will build upon this usage and start working with links and multiple pages of content.

2
Working with jQuery Mobile Pages

In the previous chapter you saw how easy it was to add jQuery Mobile to a simple HTML page. While it would be nice if every website consisted of one and only one page, real websites consist of multiple pages connected via links. jQuery Mobile makes it easy to work with multiple pages, and provides many different ways to create and link the pages.

In this chapter, we will perform the following steps:

- Add multiple pages to one jQuery Mobile file
- Discuss how links are modified by jQuery Mobile (and how to disable it)
- Demonstrate how additional files can be linked to and added to a jQuery Mobile site
- Discuss how jQuery Mobile automatically handles URLs to allow for easy bookmarking

Important preliminary points

As mentioned in the previous chapter, all of the code from this chapter is available via the ZIP file downloaded at GitHub.

Starting with this chapter, we will be presenting only the most relevant parts of each code snippet. The first code snippet, typically, will include all the code, while later code snippets will focus on the important sections. Be sure to reference the complete code snippets available via the downloaded ZIP file.

Adding multiple pages to one file

In the previous chapter, we worked on a file that had a simple page of text. For our first modification, we're going to add another page to the file and create a link to it. If you remember, jQuery Mobile looks for a particular `<div>` wrapper to help it know where your page is: `<div data-role="page">`. What makes jQuery Mobile so simple to use is that we can add another page by simply adding another div using the same format. The following code snippet `code 2-1` shows a simple example of this feature:

```
code 2-1: test1.html
<!DOCTYPE html>
<html>
  <head>
    <meta name="viewport" content="width=device-width, initial-
      scale=1">
    <title>Multi Page Example</title>
    <link rel="stylesheet"
      href="http://code.jquery.com/mobile/1.3.2/jquery.mobile-
      1.3.2.min.css" />
    <script src="http://code.jquery.com/jquery-
      1.9.1.min.js"></script>
    <script.src="http://code.jquery.com/mobile/1.3.2/jquery.mobile-
1.3.2.min.js"></script>
  </head>
  <body>
    <div data-role="page" id="homePage">
      <div data-role="header"><h1>Welcome</h1></div>
      <div data-role="content">
        <p>
          Welcome to our first mobile web site. It's going to be the
            best site you've ever seen. Once we get some content. And
            a business plan. But the hard part is done!
        </p>
        <p>
          You can also <a href="#aboutPage">learn more</a> about
            Megacorp.
        </p>
      </div>
      <div data-role="footer">
```

```
        <h4>Copyright Megacorp &copy; 2013</h4>
      </div>
    </div>
    <div data-role="page" id="aboutPage">
      <div data-role="header"><h1>About Megacorp</h1></div>
      <div data-role="content">
        <p>
          This text talks about Megacorp and how interesting it is.
            Most likely though you want to
          <a href="#homePage">return</a> to the home page.
        </p>
      </div>
      <div data-role="footer">
        <h4>Copyright Megacorp &copy; 2013/h4>
      </div>
    </div>
  </body>
</html>
```

Ok, so as always, we begin our template with a few required bits: the HTML5 DOCTYPE, the meta tag, one CSS include, and two JavaScript files. This was covered in the previous chapter and we will not be mentioning it again. Note that this template switches over to the CDN version of the CSS and JavaScript libraries:

```
<link rel="stylesheet" href="http://code.jquery.com/
  mobile/1.3.2/jquery.mobile-1.3.2.min.css" />
<script src="http://code.jquery.com/jquery-1.9.1.min.js"></script>
<script src="http://code.jquery.com/mobile/1.3.2/
  jquery.mobile-1.3.2.min.js"></script>
```

These versions are hosted by the jQuery team. Most likely, your visitors will have loaded these libraries already so they exist in their cache before arriving at your mobile site. While this is the route we will take going further with our examples, remember that you can always use the version you downloaded instead.

Notice now we have two <div> blocks. The first hasn't much changed from the previous example. We've added a unique ID (homepage), as well as a second paragraph. Notice the link in the second paragraph. It's using a standard internal link (#aboutPage) to tell the browser that we want to simply scroll the browser down to that part of the page. The target specified, aboutPage, is defined right below in another div block.

In a traditional web page, this would display as two main blocks of text on a page. Clicking any of the two links would simply scroll the browser up and down accordingly. However, jQuery Mobile is going to do something significantly different here. The following figure shows how the page is rendered in the mobile browser:

Notice something? Even though our HTML included two blocks of text (the two `<div>` blocks), it only rendered one. jQuery Mobile will always display the first page it finds, and only that page. Here comes the best part. If you click on the link, the second page automatically loads. Using your device's **back** button or simply clicking on the link will return you back to the first page. (Obviously this only works on devices that have a back button, for example Android devices.) You will also notice a smooth transition. This is something you can configure later on. But all of the interactions here, the showing and hiding of pages, and the transitions, were all done automatically by jQuery Mobile. Now is a good time to talk about links and what jQuery Mobile does when you click on them.

jQuery Mobile, links, and you

When jQuery Mobile encounters a simple link (`<a href="something.html">Foo</a>`), it will automatically capture any clicks on that link and change it to an AJAX-based load. This means that if it detects that the target is something on the same page, that is, the hash-mark style (`href="#foo"`) links we used previously, it will handle transitioning the user to a new page. If it detects a page to another file on the same server, it will use AJAX to load the page and replace the currently visible one.

If you link to an external site, then jQuery Mobile will leave the link as it is, and the normal link behavior will occur. There may be times when you want to disable jQuery Mobile from doing anything with your links at all. In that case, you can make use of a data attribute that lets the framework know it shouldn't do anything at all. An example:

```
<a href="foo.html" data-ajax="false">Normal, non-special link</a>
```

As we saw in *Chapter 1, Preparing Your First jQuery Mobile Project*, jQuery Mobile makes heavy use of data attributes. It is also very good at letting you disable behaviors you don't like. As we continue in the book you will see example after example of something jQuery Mobile does to enhance your site for mobile devices. In all of these cases though, the framework recognizes there may be times when you want to disable that.

Working with multiple files

In an ideal world, we could build an entire website with one file, never have to perform revisions, and be done with every project by 2 P.M. on Friday. But in the real world we have to deal with lots of files, lots of revisions, and, unfortunately, lots of work. In the earlier code snippet, you saw how we could include two pages within one file. jQuery Mobile handles this easily enough. But you can imagine that this would get unwieldy after a while. While we could include ten, twenty, even thirty pages, this is going to make the file difficult to work with and make the initial download for the user that much slower.

To work with multiple pages, and files, all we need to do is make a simple link to other files in the same domain as our first file. We can even combine the first technique (two pages in one file) with links to other files. In code 2-2, we've modified the first example to add a link to a new page. (As mentioned previously, we are only listing the relevant portion of the page!)

```
code 2-2:test2.html
<div data-role="page" id="homePage">
  <div data-role="header"><h1>Welcome</h1></div>
  <div data-role="content">
    <p>
      Welcome to our first mobile web site. It's going to be the
        best site you've ever seen. Once we get some content. And
        a business plan. But the hard part is done!
    </p>
    <p>
      Find out about our wonderful
      <a href="products.html">products</a>.
```

```
      </p>
      <p>
        You can also <a href="#aboutPage">learn more</a> about
          Megacorp.
      </p>
    </div>
    <div data-role="footer">
      <h4>Copyright Megacorp &copy; 2013</h4>
    </div>
  </div>
```

Now, let's look at code 2-3, our products page:

```
code 2-3: products.html
<!DOCTYPE html>
<html>
  <head>
    <meta name="viewport" content="width=device-width, initial-
      scale=1">
    <title>Products</title>
    <link rel="stylesheet"
      href="http://code.jquery.com/mobile/1.3.2/jquery.mobile-
      1.3.2.min.css" />
    <script src="http://code.jquery.com/jquery-
      1.9.1.min.js"></script>
    <script
      src="http://code.jquery.com/mobile/1.3.2/jquery.mobile-
      1.3.2.min.js"></script>
  </head>

<body>
  <div data-role="page" id="productsPage">
    <div data-role="header"><h1>Products</h1></div>
    <div data-role="content">
      <p>
        Our products include:
      </p>
      <ul>
        <li>Alpha Series</li>
        <li>Beta Series</li>
        <li>Gamma Series</li>
      </ul>
    </div>
  </div>
</body>
</html>
```

Our products page is rather simple, but notice that we included the jQuery and jQuery Mobile resources on top. Why? I mentioned earlier that jQuery Mobile is going to use AJAX to load in your additional pages. If you open up `test2.html` in any modern browser you can see for yourself using developer tools. Clicking on the link for products will fire an XHR (XHR means XML HTTP Request, but generally just means an AJAX call) request, as shown in the following screenshot from Chrome's DevTools:

⊗	◁▷ Elements	🗐 Resources	⦿ Network	🔍 Sources	✦ Timeline		
Name Path			**Method**	**Status Text**	**Type**	**Initiator**	
products.html /jqm/c2			GET	200 OK	text/html	jquery-1.9.1.min Script	

That's neat! But what happens when someone bookmarks the application? Let's now take a look at how jQuery Mobile handles URLs and navigation.

What are browser developer tools?

All modern browsers have built-in tools to help you build web pages. These tools allow you to inspect and manipulate the DOM, pause and debug JavaScript execution, and view network activity and errors.

jQuery Mobile and URLs

If you've opened up `test2.html` in your browser and played with it, you may have noticed something interesting about the URLs as you navigate. Following is the initial URL (the address and folder will, of course, differ on your computer): `http://localhost/mobile/c2/test2.html`.

After clicking on products, the URL changes to `http://localhost/mobile/c2/products.html`. If I click on **back**, and click on **learn more**, I get `http://localhost/mobile/c2/test2.html#aboutPage`.

In both subpages (the **Products** page and the **About** page) the URL was changed by the framework itself. The framework uses `history.pushState` and `history.replaceState` in browsers that support it. For older browsers, or browsers that don't support JavaScript manipulation of the URL, hash-based navigation is used instead. The products link, when viewed in an older Internet Explorer, looks like the following:

`http://localhost/mobile/c2/test2.html#/mobile/c2/products.html`

What's interesting is that in this bookmark style, `test2.html` is always loaded first. You could actually build your `products.html` to only include the `div` and be assured that if the request was made for products first, it would still render correctly. It's the newer, fancier browsers that have an issue. If you didn't include the proper jQuery and jQuery Mobile additions, when they go directly to `products.html` you would end up with a page that has no styles. It's best to simply always include your proper header files (the CSS, the JavaScript, and so on). Any decent editor will provide simple ways to create templates.

Additional customization

Working with multiple pages in jQuery Mobile is pretty simple. You could take what's been discussed in the first two chapters and build a pretty simple, but mobile compliant website right now. The following are a few more interesting tricks you may want to consider.

Page titles

You may have noticed when you clicked on the **Products** page in the previous example, the title of the browser correctly updated to `Products`. This is because jQuery Mobile noticed, and parsed in, the title tag from the `products.html` file. If you click the **learn more** link, you will notice the title also updates. How did that work? When the **About** page was loaded, jQuery Mobile used the header tag's content (`About Megacorp`) for a title. You can override this by providing an additional argument to your `div` tag defining your page: `data-title`. The following code snippet demonstrates this feature:

```
<div data-role="page" id="aboutPage" data-title="All About
Megacorp">
  <div data-role="header"><h1>About Megacorp</h1></div>
```

You can find this version in `test3.html`.

Prefetching content

The benefit of including all your content within one HTML file is that all of your pages are available immediately. But the negatives (too difficult to update, too slow for an initial download) far outweigh that. Most jQuery Mobile applications will include multiple files, and typically just use one or two pages per file. You can, however, ensure speedier loading of some pages to help improve the user experience. Imagine our **Megacorp** site. It's got three pages, but the **Products** page is a separate HTML file. Since it's the only real content on the site, most likely all of our users will end up clicking that link. We can tell jQuery Mobile to prefetch the content immediately upon the main page loading. That way, when the user does click the link, the page will load much quicker. Once again, this comes down to one simple data attribute.

```
<p>
    Find out about our wonderful <a href="products.html" data-
      prefetch="true">products</a>.
</p>
```

In the previous link, all we've done is added `data-prefetch="true"` to the link. When jQuery Mobile finds this in a link, it will automatically fetch the content right away. Now, when the user clicks the **Products** link, they will see the content even quicker. This modification was saved in `test4.html`.

Obviously, this technique should be used with care. Given a page with four main links, you may want to consider only prefetching the two most popular pages, not all four.

Changing page transitions

Earlier, we mentioned that you could configure the transitions jQuery Mobile uses between pages. Later in the book, we'll discuss how to do that globally, but if you want to switch to a different transition for a particular link, just include a `data-transition` attribute in your link:

```
<p>
Find out about our wonderful <a href="products.html" data-
  transition="pop">products</a>.
</p>
```

Valid values for transition include: `fade` (the default), `flip`, `flow`, `pop`, `slide`, `slidedown`, `slidefade`, `turn`, and `none`.

Many transitions also support a reverse action. Normally jQuery Mobile figures out if you need this, but if you want to force a direction, use the data-direction attribute:

```
<p>
Find out about our wonderful <a href="products.html" data-
  transition="pop" data-direction="reverse">products</a>.
</p>
```

Summary

This chapter further fleshed out the concept of jQuery Mobile pages, and how to work with multiple pages. Specifically, we saw how one physical file can contain many different pages. jQuery Mobile will handle hiding all but the first page. We also saw how you can link to other pages, and how jQuery Mobile uses AJAX to dynamically load the content into the browser. Next, we discussed how jQuery Mobile handles updating the URL of the browser in order to enable bookmarking. Finally, we discussed two utilities that will help to improve your page. The first way was to provide a title for embedded pages. The second technique demonstrated how to prefetch content to further improve the experience of the users visiting your site.

In the next chapter, we'll take a look at headers, footers, and navigation bars. These will greatly enhance our pages and make them easier to navigate.

3
Enhancing Pages with Headers, Footers, and Toolbars

Toolbars provide a simple way to add navigation elements to a mobile website. They can be especially useful for consistent or site-wide navigation controls that users can always refer to when navigating through your application.

In this chapter, we will perform the following steps:

- Talk about how to create both headers and footers
- Discuss how to turn these headers and footers into useful toolbars
- Demonstrate how to create fixed positioned toolbars that always show up, no matter how large the content of a particular page is
- Show an example of navigation bars

Important preliminary points

As mentioned in the previous chapter, all of the code from this chapter is available via the ZIP file downloaded at GitHub. As before, the chapter will only consider the important parts of each file. Consult the downloaded files for the complete source.

Adding headers

You've already worked with headers in the previous chapter, so the code will be familiar. In this chapter, we will study them deeper and demonstrate how to add additional functionality, such as buttons, to your site headers.

If you remember, a header can be defined by simply using a div with the appropriate role:

```
<div data-role="header"><h1>My Header</h1></div>
```

We can further add functionality to headers by adding buttons. Buttons could be used for navigation (for example, to return to the home screen), or to provide links to related pages. Because the center of the header is used for text, there are only two spaces available for buttons—one to the left and one to the right. Buttons can be added simply by creating links in your header. The first link will be to the left of the text and the second link to the right. The following file is an example:

```
code 3-1: header_test.html
<div data-role="header">
  <a href="index.html">Home</a>
  <h1>My Header</h1>
  <a href="contact.html">Contact</a>
</div>
```

When viewed in the mobile browser, you can see that the links were automatically turned into buttons:

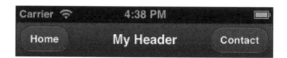

Notice how the simpler links were automatically turned into big buttons, making them easier to use and more "control-like" for the header. You may be wondering, what if you only want one button, and want it on the right-hand side? Removing the first button and keeping the second in place will not work, as shown in the following code snippet:

```
<div data-role="header">
  <h1>My Header</h1>
  <a href="contact.html">Contact</a>
</div>
```

The previous code snippet creates a button in the header but on the left-hand side. In order to position the button to the right, simply add the class ui-btn-right.

```
code 3-2: header_test2.html
<div data-role="header">
  <h1>My Header</h1>
  <a href="contact.html" class="ui-btn-right">Contact</a>
</div>
```

You can also specify `ui-btn-left` to place a link on the left-hand side, but as demonstrated in the previous code snippet, that's the normal behavior.

Icon sneak peak

While not specifically a header toolbar feature, one interesting feature available to all buttons in jQuery Mobile is the ability to specify an icon. A set of simple, easily recognizable icons ship with jQuery Mobile, and are available to use immediately. These icons will be discussed further in *Chapter 6, Working with Forms and jQuery Mobile*, but as a quick preview, the following code snippet shows a header with two customized icons:

```
code 3-3: header_test3.html
<div data-role="header">
  <a href="index.html" data-icon="home">Home</a>
  <h1>My Header</h1>
  <a href="contact.html" data-icon="info">Contact</a>
</div>
```

Notice the new attribute, `data-icon`. When viewed in the browser, you get what is shown in the following screenshot:

The specific icons displayed were based on the values passed to the `data-icon` attributes. Again, this will be discussed more in depth later in the book.

Working with back buttons

Depending on your user's hardware, they may or may not have a physical back button. For devices that do, such as Android phones, hitting the back button will work just fine in a jQuery Mobile application. Whatever page the user was on previously will be loaded as soon as the button is clicked. But on other devices, like the iPhone, there is no such button to click. While you can provide links to navigate around pages yourself, jQuery Mobile provides some nice built in support for navigating backwards out of the box.

There are two ways you can add an automatic back button. code 3-4 shows a simple, two page jQuery Mobile site. In the second page, we've added a new data attribute, `data-add-back-btn="true"`. This will create a back button in the header of the second page automatically. Next, we also added a simple link in the page content. While the actual URL for the link is blank, make note of the `data-rel="back"` attribute. jQuery Mobile will detect this link and automatically send the user to the previous page.

```
code 3-4: back_button_test.html
<div data-role="page">

  <div data-role="header"><h1>My Header</h1></div>

  <div data-role="content">
    <p>
      <a href="#subpage">Go to the sub page...</a>
    </p>
  </div>

</div>


<div data-role="page" id="subpage" data-add-back-btn="true">
  <div data-role="header"><h1>Sub Page</h1></div>
  <div data-role="content">
    <p>
      <a href="" data-rel="back">Go back...</a>
    </p>
  </div>
</div>
```

The following screenshot demonstrates the feature in action:

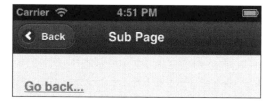

In case you're curious, the text of the button can be customized by simply using another data attribute in your page div: `data-add-back-btn="true" data-back-btn-text="Return"`. You can turn on back button support globally and change the text via JavaScript as well. This will be discussed in *Chapter 9, jQuery Mobile Configuration, Utilities, and JavaScript Methods*.

As a final example, what if you want to create a header without any actual text? Imagine for a moment that you include this header:

```
<div data-role="header">
  <a href="index.html" data-icon="home">Home</a>
</div>
```

When viewed in the mobile browser, the header is not properly sized because it is missing the `<h1>` normally used to provide text.

Luckily, there is a simple enough fix for this. Add a span with the class `ui-title` to the header and all will be fine:

```
<div data-role="header">
  <a href="index.html" data-icon="home">Home</a>
  <span class="ui-title"></span>
</div>
```

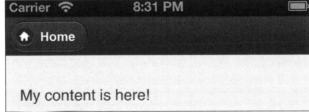

Working with footers

Footers are going to be, for the most part, much like headers. We've previously demonstrated the use of the `data-role` to create a footer:

```
<div data-role="footer"><h4>My Footer</h4></div>
```

As with the headers, you can include buttons in the footer. Unlike the headers, the buttons in a footer do not automatically position themselves to the left and right of the text. Rather, they simply line up from the left-hand side. The following is a simple example with two buttons:

```
<div data-role="footer">
  <a href="credits.html">Credits</a>
  <a href="contact.html">Contact</a>
</div>
```

The following screenshot demonstrates this feature:

This works, but notice the buttons don't have much space around them. You can improve that by adding a class called ui-bar to your footer div tag, as shown in the following code snippet:

```
<div data-role="footer" class="ui-bar">
  <a href="credits.html">Credits</a>
  <a href="contact.html">Contact</a>
</div>
```

You can find both of the previous examples in the files footer_test2.html and footer_test3.html.

If you include footer text along with your buttons, you should *not* use an <h4> tag around the text. This is a bit different from headers and can trip you up if you forget. If you do forget, your header will end up approximately 3 times as high as it needs to be!

Creating fixed and full-screen headers and footers

In the previous discussion about headers and footers, you saw a few examples of how buttons can be added. These buttons could be useful for navigating in your site. But what if a particular page is somewhat long? A blog entry, for example, could be quite long, especially when viewed on a mobile device. As the user scrolls, the header or footer could be off-screen. jQuery Mobile provides a way to create fixed position headers and footers.

With this feature enabled, the header and footer will always be visible. In a page with long content, the user can scroll up and down, but the headers and footers will remain in their proper positions. This only works with mobile browsers that support the fixed value for the CSS position property. For browsers that do not support this feature, the headers and footers will act as normal. This feature can be enabled by adding data-position="fixed" to the div tag used for either the header or footer. code 3-5 demonstrates an example. In order to ensure the page actually scrolls, many paragraphs of text were repeated. This has been removed from the code in the book, but exists in the actual file.

```
code 3-5: longpage.html
<div data-role="page">
  <div data-role="header" data-position="fixed"><h1>My
    Header</h1></div>

  <div data-role="content">
    <p>
      Lorem ipsum dolor sit amet, consectetur adipiscing elit.
        Suspendisse id posuere lacus. Nulla ac sem ut eros
        dignissim interdum a et erat. Class aptent taciti
        sociosqu ad litora torquent per conubia nostra, per
        inceptos himenaeos. In ac tellus est. Nunc consequat

        parturient montes, nascetur ridiculus mus. In id volutpat
        lectus.Quisque mauris ipsum, vehicula id ornare aliquet,
        auctor volutpat dui. Sed euismod sem in arcu dapibus
        condimentum dictum nibh consequat.
    </p>
```

```
    </div>

    <div data-role="footer" data-position="fixed"><h4>My
       Footer</h4></div>
    </div>
```

We won't bother with a screenshot of this example as it won't exactly convey the feature well. But if you try this in your mobile device, notice while scrolling up and down, that as soon as you lift your finger the header and footer will both pop in. This gives the user access to them no matter how large the page may be.

Full-screen headers and footers

Another option to consider is what's called full-screen headers and footers. This is a metaphor commonly used with pictures, but can also be used where fixed-positioned headers and footers are used. In this scenario, the header and footer appear and disappear with clicks. So, with a photo, this allows you a view of the photo as it is, but also the ability to get the header and footer back with a simple click. Perhaps, instead of full-screen headers or footers, you can consider them as retrievable headers and footers instead. In general, this is best used when you want the content of the page to be viewed by itself—again, an excellent example of this would be pictures.

To enable this feature, simply add data-fullscreen="true" to the header and footer tags in a page. (Yes, you can choose to enable full-screen support for only the header or footer, if you wish.) code 3-3 demonstrates this feature, as shown in the following code snippet:

```
code 3-6: fullscreen.html
<div data-role="page">
  <div data-role="header" data-position="fixed" data-
    fullscreen="true"><h1>My Header</h1></div>
  <div data-role="content">
    <p>
      <img src="green.png" title="Green Block">
    </p>
  </div>
  <div data-role="footer" data-position="fixed" data-
    fullscreen="true"><h4>My Footer</h4></div>
</div>
```

As with the previous example, the code snippet doesn't translate very well to static screenshots. Open it up in your mobile browser and take a look. Remember, you can click multiple times to toggle the effect on and off.

Working with navigation bars

You've now seen a few examples of headers and footers that include buttons, but jQuery Mobile has a cleaner version of this called **NavBars** (**navigation bars**). These are full-screen-wide bars used to hold buttons. jQuery Mobile also supports highlighting one button at a time as an active button. When used for navigation, this is an easy way to mark a page as being active.

A NavBar is simply an unordered list wrapped in a `div` tag that uses `data-role="navbar"`. Placed inside a footer it looks similar to the following code snippet:

```
<div data-role="footer">
  <div data-role="navbar">
    <ul>
      <li><a href="header_and_footer__with_navbar.html" class="ui-
        btn-active">Home</a></li>
      <li><a href=" header_and_footer__with_navbar_credits.html "
        >Credits</a></li>
      <li><a href=" header_and_footer_ with_navbar_contact.html "
        >Contact</a></li>
    </ul>
  </div>
</div>
```

Notice the use of `class="ui-btn-active"` on the first link. This will mark the first button as active. jQuery Mobile won't be able to do this for you automatically, so as you build each page and make use of a `navbar` you will have to move the class appropriately. The following screenshot shows how it looks:

You can add up to 5 buttons and jQuery Mobile will appropriately size the buttons to make them fit. If you go over five, then the buttons will simply be split over multiple lines. Most likely, this is not something you want to cover. Overwhelming the user with too many buttons is a sure way to confuse, and ultimately anger your users.

You can also include a `navbar` in your header. If placed after the text, or any other buttons, jQuery Mobile will automatically drop it to the next line:

```
<div data-role="header">
  <h1>Home</h1>
  <div data-role="navbar">
    <ul>
      <li><a href=" header_and_footer_with_navbar.html" class="ui-btn-
        active">Home</a></li>
      <li><a href=" header_and_footer_with_navbar_credits.html"
        >Credits</a></li>
      <li><a href=" header_and_footer_with_navbar_contact.html"
>Contact</a></li>
    </ul>
  </div>
</div>
```

You can see an example of both of these in action in the file named `header_and_footer_with_navbar.html`.

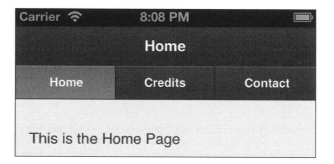

Persisting navigation bar footers across multiple pages

Let's now take two of the previous topics and combine them into one incredibly cool little feature—multiple page persistent footers. It's a bit more work, but you can create a footer, NavBar, that will not disappear when switching from page to page. In order to do this, you have to follow a few simple rules:

- Your footer `div` must be present on all pages
- Your footer `div` must use the same `data-id` value across all pages
- You must use two CSS classes, `ui-state-persist` and `ui-btn-active`, on the active page in the NavBar
- You must also use the persistent footer feature

That sounded a bit complex, but it's really just a tiny bit more HTML in your template. In code 3-7, an index page for a fictional company makes use of a footer, NavBar. Note the use of `ui-state-persist` and `ui-btn-active` for the currently selected page.

```
code 3-7: persistent_footer_index.html

<div data-role="footer" data-position="fixed" data-id="footernav">
  <div data-role="navbar">
    <ul>
      <li><a href="persistent_footer_index.html" class="ui-btn-
        active ui-state-persist">Home</a></li>
      <li><a
        href="persistent_footer_credits.html">Credits</a></li>
      <li><a
        href="persistent_footer_contact.html">Contact</a></li>
    </ul>
  </div>
</div>
```

The following screenshot shows how the complete page looks:

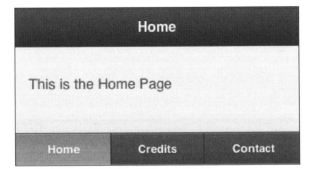

We don't need to worry so much about the other two pages. You can find them in the ZIP file you downloaded. The following code snippet is the footer section from the second page. Notice that the only change here is the movement of the `ui-btn-active` and `ui-state-persist` class:

```
<div data-role="footer" data-position="fixed" data-id="footernav">
  <div data-role="navbar">
    <ul>
      <li><a href="persistent_footer_index.html">Home</a></li>
      <li><a href="persistent_footer_credits.html" class="ui-btn-
        active ui-state-persist">Credits</a></li>
      <li><a
```

```
            href="persistent_footer_contact.html">Contact</a></li>
      </ul>
    </div>
  </div>
```

Clicking from one page to another shows a smooth transition to each page, but the footer bar remains. Much like a framed site (don't shudder—frames weren't always looked at with scorn), the footer will stay as the user navigates throughout the site.

Summary

In this chapter, we discussed how to add headers, footers, and navigation bars (NavBars) to your jQuery Mobile pages. We showed how the proper `div` tags will create nicely formatted headers and footers on your page, and how to make these headers and footers persist over a long page. furthermore, we demonstrated full-screen mode for headers and footers. These are headers and footers that appear and disappear with clicks—perfect for images and other items you want to show in a full-screen type view on your mobile device. Finally, we saw how to combine persistent footers and NavBars to create a footer that doesn't go away when the page changes.

In the next chapter, we'll take an in-depth look at lists. Lists are one of the primary ways folks add navigation and menus to their mobile sites. jQuery Mobile provides a plethora of options for creating and styling lists.

4
Working with Lists

Lists are a great way to provide menus to users on a mobile website. jQuery Mobile provides a wealth of list options, from simple lists to lists with custom thumbnails and multiple user actions.

In this chapter, we will cover the following aspects:

- Talk about how to create lists
- How to create linked and sub-menu style lists
- How to create different styles of lists

Creating lists

As you've (hopefully!) come to learn, jQuery Mobile takes an approach of enhancement when it comes to UI. You take the ordinary, simple HTML, add a bit of markup (sometimes!), and jQuery Mobile will do the heavy lifting of enhancing the UI. The same process applies to lists. We've all worked with simple lists in HTML before; the following code snippet is an example:

```
<ul>
  <li>Raymond Camden</li>
  <li>Scott Stroz</li>
  <li>Todd Sharp</li>
  <li>Dave Ferguson</li>
</ul>
```

And we all know how they are displayed (a bulleted list in the case of the previous code snippet). Let's take that list and drop it in a simple jQuery Mobile optimized page. code 4-1 takes a typical page and drops in our list:

```
code 4-1: test1.html
<div data-role="page">
  <div data-role="header">
    <h1>My Header</h1>
  </div>
  <div data-role="content">
    <ul>
      <li>Raymond Camden</li>
      <li>Scott Stroz</li>
      <li>Todd Sharp</li>
      <li>Dave Ferguson</li>
    </ul>
  </div>
  <div data-role="footer">
    <h1>My Footer</h1>
  </div>
</div>
```

Given this HTML, jQuery Mobile gives us something nice right away, as shown in the following screenshot:

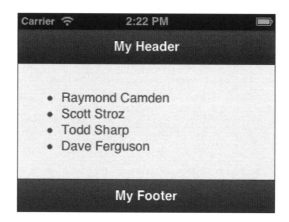

We can enhance that list, though, with a simple change. Take the ordinary `<ul>` tag from code 4-1, and add a data-role="listview" attribute, as shown in the following line of code:

```
<ul data-role="listview">
```

In the code you download from GitHub, you can find this modification in test2. html. The change, though, is rather dramatic, as shown in the following screenshot:

You can see that the items no longer have the bullets in front, but they are much larger and easier to read. Things get even more interesting when we begin to add links to our list. In the following code snippet I've added a link to each list item:

```
<ul data-role="listview">
   <li><a href="ray.html">Raymond Camden</a></li>
   <li><a href="scott.html">Scott Stroz</a></li>
   <li><a href="todd.html">Todd Sharp</a></li>
   <li><a href="dave.html">Dave Ferguson</a></li>
</ul>
```

Once again you can find the complete file; this was taken from the ZIP file you downloaded earlier. This one may be found in test3.html. The following screenshot shows how this code is rendered:

Notice the new arrow image. This was automatically added when jQuery Mobile detected a link in your list. Now you've turned a relatively simple HTML unordered list into a simple menu system. This, by itself, is pretty impressive, but as we will see throughout the remaining chapter, jQuery Mobile provides a wealth of rendering options to let you customize your lists.

Working with list features

jQuery Mobile provides multiple different styles of lists, as well as different features that can be applied to them. For the next part of this chapter we'll cover some (but not all!) of these options available. These aren't in any particular order and are presented as a gallery of options available to you. You probably will not (and should not!) try to use all of these within one application, but it's good to keep in mind the various types of list styles jQuery Mobile has available.

Creating inset lists

One of the simplest and slickest changes you can make to your lists is to turn them into **inset lists**. These are lists that do not take up the full width of the device. Taking the initial list we modified with `data-role="content"`, we can simply add another attribute, `data-inset="true"`, for the following code block (found in `test4.html`):

```
<ul data-role="listview" data-inset="true">
  <li>Raymond Camden</li>
  <li>Scott Stroz</li>
  <li>Todd Sharp</li>
  <li>Dave Ferguson</li>
</ul>
```

The result is now very different from the earlier example:

Creating list dividers

Another interesting UI element you may wish to add to your lists are dividers. These are a great way to separate a long list into something that is a bit easier to scan. Adding a list divider is as simple as adding a li tag that makes use of data-role="list-divider". The following code snippet shows a simple example of this element:

```
<ul data-role="listview" data-inset="true">
  <li data-role="list-divider">Active</li>
  <li>Raymond Camden</li>
  <li>Scott Stroz</li>
  <li>Todd Sharp</li>
  <li data-role="list-divider">Archived</li>
  <li>Dave Ferguson</li>
</ul>
```

In the previous code block, note the two new li tags making use of the list-divider role. In this example, I've used these to separate the list of people into two groups. You can find the complete template in test5.html. The following screenshot shows how this is rendered:

Autodividers

Another option for dividers is to have jQuery Mobile create them for you automatically. By adding `data-autodividers="true"` to any listview, jQuery Mobile will automatically divide lists by the first letter of the item. (You can tweak this with custom JavaScript if you wish.) Given the following simple list (from `test6.html`):

```
<ul data-role="listview" data-inset="true" data-
   autodividers="true">
   <li>Apples</li>
   <li>Apricots</li>
   <li>Bananas</li>
   <li>Cherries</li>
   <li>Coconuts</li>
</ul>
```

Notice that there are no dividers in the list, but that autodividers is enabled at the top level of the list itself. This creates the following in the browser:

If you were to add a new list item—let's say for donuts—`listview` will automatically add a new D divider. And let's be honest, no list is complete without donuts.

Creating lists with count bubbles

Yet another interesting UI trick you can perform with jQuery Mobile lists are count bubbles. This is a UI enhancement that adds a simple number to the end of each list item. The numbers are wrapped in a bubble-like look, which is commonly used for e-mail-style interfaces. In the following code snippet, the count bubble is used to signify the number of cookies consumed at a technical conference. This number is supplied inside a span tag that uses a class of ui-li-count.

```
<ul data-role="listview" data-inset="true">
  <li data-role="list-divider">Cookies Eaten</li>
  <li>Raymond Camden <span class="ui-li-count">9</span></li>
  <li>Scott Stroz <span class="ui-li-count">4</span></li>
  <li>Todd Sharp <span class="ui-li-count">13</span></li>
  <li>Dave Ferguson <span class="ui-li-count">8</span></li>
</ul>
```

A simple HTML change is demonstrated—but consider how nicely it gets rendered—as shown in the following screenshot:

You can find a complete example of this feature in test7.html.

Using thumbnails and icons

Another common need with lists is to include images. jQuery Mobile supports both thumbnails (smallish images) and icons (even smaller images) that display well within the list control. Let's first look at including thumbnails within your list. Assuming you already have nicely sized images (our examples are all 160 pixels wide by 160 pixels high), you can simply include them within each li element as demonstrated in the following code snippet:

```
<ul data-role="listview" data-inset="true">
  <li><a href="ray.html"><img src="ray.png"> Raymond
    Camden</a></li>
  <li><a href="scott.html"><img src="scott.png"> Scott
    Stroz</a></li>
  <li><a href="todd.html"><img src="todd.png"> Todd Sharp</a></li>
  <li><a href="dave.html"><img src="dave.png"> Dave
    Ferguson</a></li>
</ul>
```

Nothing special is done with the image, no data attribute or class is added. jQuery Mobile will automatically left align the image, and place the item text aligned to the top of each `li` block as shown in the following screenshot:

You can find the previous demonstration in `test8.html`. So what about icons? To include an icon in your code, add the class `ui-li-icon` to your image. (Note that the beginning of the class is `ui`, not `ul`.) The following code snippet is an example of that with our same list:

```
<ul data-role="listview" data-inset="true">
  <li>
    <a href="ray.html">
      <img src="ray_small.png" class="ui-li-icon">
      Raymond Camden
    </a>
  </li>
  <li>
    <a href="scott.html">
      <img src="scott_small.png" class="ui-li-icon">
      Scott Stroz
    </a>
  </li>
  <li>
    <a href="todd.html">
      <img src="todd_small.png" class="ui-li-icon">
      Todd Sharp
```

```
    </a>
  </li>
  <li>
    <a href="dave.html">
      <img src="dave_small.png" class="ui-li-icon">
      Dave Ferguson
    </a>
  </li>
</ul>
```

jQuery Mobile does shrink images when used with this class, but in my experience, the formatting was better when the image was resized beforehand. Doing so also improves the speed of your web page as the smaller images should result in quicker download times. The images previously are all 16 pixels wide and high each. And the result is as follows:

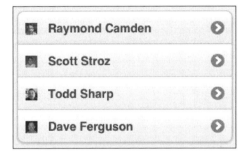

You can find the previous example in `test9.html`.

Creating split button lists

Another interesting feature of jQuery Mobile lists is the split button list. This is simply a list with multiple actions. A main action is activated when the user clicks on the list item and a secondary action is available via a button at the end of the list item. For this example, let's start with the screenshot first and then demonstrate how it's done:

As you can see, each list item has a secondary icon at the end of the row. This is an example of a split item list and is defined by simply adding a second link to a list item. For example:

```
<ul data-role="listview" data-inset="true">
  <li>
    <a href="ray.html">
      <img src="ray_small.png" class="ui-li-icon">
      Raymond Camden
    </a>
    <a href="foo.html">Delete</a>
    </li>
  <li>
    <a href="scott.html">
      <img src="scott_small.png" class="ui-li-icon">
      Scott Stroz
    </a>
    <a href="foo.html">Delete</a>
  </li>
  <li>
    <a href="todd.html">
      <img src="todd_small.png" class="ui-li-icon">
      Todd Sharp
    </a>
    <a href="foo.html">Delete</a></li>
  <li>
    <a href="dave.html">
      <img src="dave_small.png" class="ui-li-icon">
      Dave Ferguson
    </a>
    <a href="foo.html">Delete</a>
  </li>
</ul>
```

Note that the second link's text, Delete, is actually replaced by the icon. You can specify an icon by adding the data attribute split-icon to your ul tag, as shown in the following line of code:

```
<ul data-role="listview" data-inset="true" data-split-
icon="delete">
```

The complete code for this example may be found in test10.html.

Using a search filter

For our last and final list feature we will look at the search filters. The lists we've worked with so far have been pretty short. Longer lists may make it difficult for users to find what they are looking for. jQuery Mobile provides an incredibly simple way to add a search filter to your lists. By adding `data-filter="true"` to any list, jQuery Mobile will automatically add a search field on top that filters as you type:

```
<ul data-role="listview" data-inset="true" data-filter="true">
  <li><a href="ray.html">Raymond Camden</a></li>
  <li><a href="scott.html">Scott Stroz</a></li>
  <li><a href="todd.html">Todd Sharp</a></li>
  <li><a href="dave.html">Dave Ferguson</a></li>
  (lots of items….)
</ul>
```

The result looks similar to the following screenshot:

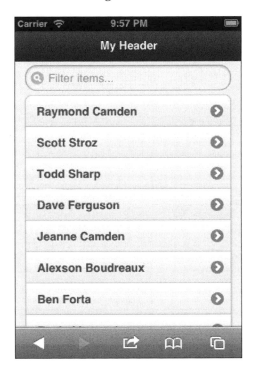

If you begin typing in the previous field, the list automatically filters out the results as you type:

By default, the search is case-insensitive and matches anywhere in the list item. You can specify placeholder text for the search form by using `data-placeholder-text="Something"` in your `ul` tag. You can also specify a theme for the form using `data-filter-theme`.

Along with filtering, you can reverse the process and create something that looks like an autocomplete control. This is done by adding `data-filter-reveal="true"` to your unordered list. Here is an example demonstrating this:

```
<ul data-role="listview" data-inset="true" data-filter="true"
data-filter-reveal="true" data-filter-placeholder="Type to
search...">
```

Note that you can also customize the text in the search field as well. Now on, display the list is completely hidden.

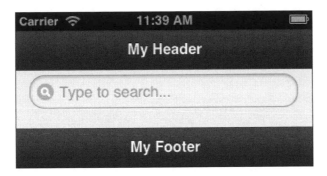

As you type, elements in the list that match your input will be displayed. You can also, via a bit of JavaScript, completely modify the behavior of this widget so as to retrieve the contents to display via an AJAX (XHR) request.

Summary

This chapter discussed how to work with listviews in jQuery Mobile. We saw how to turn a regular HTML list into a mobile optimized list, and we demonstrated the numerous types of list features available with the framework.

In the next chapter, we'll take what we've learned already and build a real (albeit a bit simple) mobile-optimized website for a hotel.

5

Getting Practical – Building a Simple Hotel Mobile Site

In the past four chapters, we've looked at a few features of jQuery Mobile, but we already have enough knowledge to build a simple, pretty basic mobile-optimized website.

In this chapter, we will cover the following aspects:

- Discuss what our hotel mobile website will contain
- Create the hotel mobile website using jQuery Mobile
- Discuss what could be done to make the site more interactive

Welcome to Hotel Camden

The Hotel Camden, known throughout the world, has had a web presence for some time now. (Ok, just to be clear, we're making this up!) They were an early innovator in the online world, beginning with a simple website in 1996 and gradually improving their online presence over the years. Online visitors to the Hotel Camden can now see virtual tours of rooms, check the grounds with a stunning 3D Adobe Flash plugin, and actually make reservations online. Recently, though, the owners of Hotel Camden have decided they want to move into the mobile space. For now, they want to start simply and create a mobile-optimized site which includes the following features:

- **Contact information**: This will include both a phone number and an e-mail address. Ideally, the user will be able to click these and get connected to a real person.
- **Map of the hotel location**: This should include the address and possibly a map too.

- **Room types available**: This can be a simple list of the rooms from the simplest to the most grand.

- **Provide a way for the user to get to the real website:** We are accepting that our mobile version will be somewhat limited (for this version), so at a minimum, we should provide a way for users to return to the desktop version of the site.

The home page

Let's begin with the initial home page for the Camden Hotel. This will provide a simple list of options, as well as a bit of marketing text on the top. The text doesn't actually help anyone, but the marketing staff won't let us release the site without it.

```
code 5-1: index.html
<!DOCTYPE html>
<html>
  <head>
    <title>The Camden Hotel</title>
    <meta name="viewport" content="width=device-width, initial-
      scale=1">
    <link rel="stylesheet" href="http://code.jquery.com/mobile/
      1.3.2/jquery.mobile-1.3.2.min.css" />
    <script src="http://code.jquery.com/jquery-
      1.9.1.min.js"></script>
    <script src="http://code.jquery.com/mobile/
      1.3.2/jquery.mobile-1.3.2.min.js"></script>
  </head>
  <body>
    <div data-role="page">
      <div data-role="header">
        <h1>Camden Hotel</h1>
      </div>
      <div data-role="content">
        <p>
          Welcome to the Camden Hotel. We are a luxury hotel
            specializing in catering to the rich and overly
            privileged. You will find our accommodations both
            flattering to your ego, as well as damaging to your
            wallet. Enjoy our complimentary wi-fi access, as well as
            caviar baths while sleeping in beds with gold thread.
        </p>
        <ul data-role="listview" data-inset="true">
          <li><a href="find.html">Find Us</a></li>
          <li><a href="rooms.html">Our Rooms</a></li>
          <li><a href="contact.html">Contact Us</a></li>
```

```
        <li><a href="">Non-Mobile Site</a></li>
      </ul>
    </div>
    <div data-role="footer">
      <h4>&copy; Camden Hotel 2012</h4>
    </div>
  </div>
  </body>
</html>
```

At a higher level, the code in `code 5-1` is simply another instance of the jQuery page model we've discussed before. You can see what the CSS and JavaScript include, as well as the `div` wrappers that set up our page, header, footer, and the content. Within our `content div` you can also see a list being used. We've left the URL blank for the non-mobile site option (`Non-Mobile Site`), as we don't have a real website for the Camden Hotel.

The order of the list items is also thought out. Each item is listed in order of what the staff feel are the most common requests, with the number one being simply finding the hotel and the last option (ignoring leaving the site) being able to contact the hotel. Over all, the idea of this example is to provide quick access to the most important aspects of what we think the hotel customers will need. The following screenshot shows how the site looks:

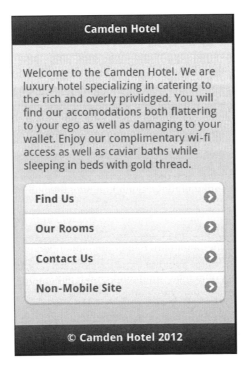

It isn't terribly exciting, but it renders well and is easy to use. Later on, you'll learn how to theme jQuery Mobile so your site doesn't look like every other example out there.

Finding the hotel

The next page of our mobile website is focused on helping the user find the hotel. This will include the address, as well as a map. code 5-2 shows how this is done (as a reminder, we are saving space and trimming the code samples a bit!):

```
code 5-2: find.html
<div data-role="page">
  <div data-role="header">
    <h1>Find Us</h1>
  </div>
  <div data-role="content">
    <p>
      The Camden Hotel is located in beautiful downtown
        Lafayette, LA. Home of the Ragin Cajuns, good food, good
        music, and all around good times, the Camden Hotel is
        smack dab in the middle of one of the most interesting
        cities in America!
    </p>
    <p>
      400 Kaliste Saloom<br/>
      Lafayette, LA<br/>
      70508
    </p>
    <p>
      <img src="http://maps.googleapis.com/maps/api/
        staticmap?center=400+Kaliste+Saloom,+Lafayette,
        LA&zoom=12&size=150x150&scale=2&maptype=roadmap&
        markers=label:H%7C400+Kaliste+Saloom,+Lafayette,
        LA&sensor=false">
    </p>
  </div>
  <div data-role="footer">
    <h4>&copy; Camden Hotel 2012</h4>
  </div>
</div>
```

The beginning of the template has our boiler plate included again, and as before, we have some marketing speak fluff on top. Immediately below this, though, is the address and the map. We created the map using one of the cooler Google features, Static Maps. You can read more about Google Static Maps at its home page: `http://code.google.com/apis/maps/documentation/staticmaps/`. Essentially, it is a way to create static maps via URL parameters. There's no zooming or panning in these maps, but if you are just trying to show a user where your business is located, it's an incredibly powerful and simple way to do so. Unlike the traditional Google Maps API, which is driven by JavaScript, the Static Maps API is simply an image URL with URL parameters specifying options for size, location, and other factors. While there are a large number of options, you can use with this API; our example simply centers it on an address and adds a marker there as well. The address used is simply one in my hometown and does not reflect a real business. The label **H** is used for the marker, but a custom icon could be used instead. The following screenshot shows how this looks:

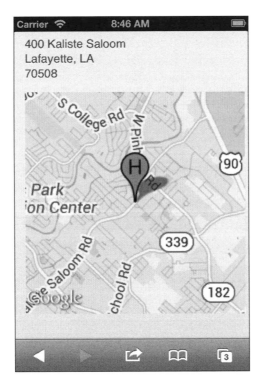

You could play around with that map URL a bit more to change the zoom, change the colors, and so on to your liking.

Listing the hotel rooms

Now let's look at `rooms.html`. This is where we will list out the room types available at the hotel:

```
code 5-3: rooms.html
<div data-role="page">
  <div data-role="header">
    <h1>Our Rooms</h1>
  </div>
  <div data-role="content">
    <p>
      Select a room below to see a picture.
    </p>
    <ul data-role="listview" data-inset="true">
      <li><a href="room_poor.html">Simple Elegance</a></li>
      <li><a href="room_medium.html">Gold Standard</a></li>
      <li><a href="room_high.html">Emperor Suite</a></li>
    </ul>
  </div>
  <div data-role="footer">
    <h4>&copy; Camden Hotel 2012</h4>
  </div>
</div>
```

The rooms page is simply a list of their rooms. The hotel has three levels of rooms, each linked to the list so that the user can get details. You can find all three files in the ZIP downloaded from GitHub, but let's look at one of them in detail:

```
code 5-4: room_high.html
<div data-role="page" data-fullscreen="true">
  <div data-role="header" data-position="fixed">
    <h1>Emperor Suite</h1>
  </div>
  <div data-role="content">
    <img src="room2.jpg" />
  </div>
  <div data-role="footer" data-position="fixed">
    <h4>&copy; Camden Hotel 2012</h4>
  </div>
</div>
```

The room detail page is only an image. Not very helpful, but it gets the point across. However, notice that we use a trick we learned in *Chapter 3, Enhancing Pages with Headers, Footers, and Toolbars*. This allows the user to quickly click and hide the headers so they can see the room in all its glory.

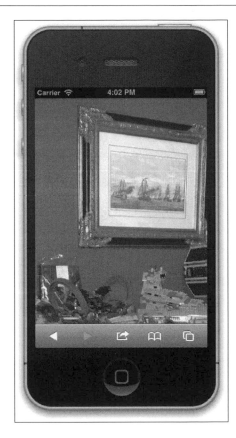

Contacting the hotel

Now, let's take a look at the contact page. This will provide the user with information on how to reach the hotel:

```
code 5-5: contact.html
<div data-role="page">
  <div data-role="header">
    <h1>Contact Us</h1>
  </div>
  <div data-role="content">
    <p>
      <b>Phone:</b> <a href="tel:555-555-5555">
        555-555-5555</a><br/>
      <b>Email:</b> <a href="mailto:people@camdenhotel.fake">
        people@camdenhotel.fake</a>
    </p>
  </div>
```

```
<div data-role="footer">
  <h4>&copy; Camden Hotel 2012</h4>
</div>
</div>
```

As before, we've wrapped our page in the proper script blocks and `div` tags. Make a special note of our two links. Both the phone and e-mail links use URLs that may not look familiar to you. The first, `tel:555-555-555`, is actually a way to ask the mobile device to call a phone number. Clicking it brings up the dialer, as shown in the following screenshot:

This makes it easy for the user to quickly call the hotel. Similarly, the `mailto` link will allow for quickly jotting an e-mail off to the hotel. Other URL schemes exist, including ones to send an SMS message. As you can probably guess, this scheme uses the form `sms`, so to begin an SMS message to a phone number, you could use the following URL: `sms://5551112222`.

For additional examples, consider the official documentation for iOS URL schemes: `http://developer.apple.com/library/ios/#featuredarticles/iPhoneURLScheme_Reference/Introduction/Introduction.html`. For a good overview of other platforms, the HTML5 Mobile Development DZone Reference Card is an excellent resource: `http://refcardz.dzone.com/refcardz/html5-mobile-development`.

Summary

In this chapter, we took what we've learned so far and built a very simple, but effective website for a fake hotel. This website shared essential information for folks needing to learn about the hotel while on a mobile device, made use of Google's Static Maps API to create a simple map showing the hotel's location, and demonstrated the use of `tel` and `mailto` URL schemes for automatic phone dialing and e-mailing.

In the next chapter, we'll take a look at forms and how they are automatically improved with jQuery Mobile.

6
Working with Forms and jQuery Mobile

In this chapter, we will look at forms. Forms are a critical part of most websites as they provide the primary way for users to interact with the website. jQuery Mobile goes a long way to making forms both usable and elegantly designed for mobile devices.

In this chapter, we will cover the following aspects:

- Talk about what jQuery Mobile does with forms
- Work with a sample form and describe how the results are handled
- Discuss specifics about how to build certain types of forms and make use of jQuery Mobile conventions

Before you begin

In this chapter, we're going to talk about forms and how jQuery Mobile enhances them. As part of our discussion, we will be posting our forms to the server. In order to have the server actually do something with the response, we're going to make use of an application server from Adobe called ColdFusion. ColdFusion is not free for production use, but is 100% free for development and is a great server for building web applications. You do not need to download ColdFusion. If you do not, the forms you use within this chapter should not be submitted. This chapter does talk about how forms are submitted, but the response to the forms isn't really critical.
If you know another language, like PHP, you should be able to simply mimic the code ColdFusion is using to echo back the form data.

ColdFusion (currently version 10) can be downloaded at `http://www.adobe.com/go/coldfusion`. Versions exist for Windows, OS X, and Linux. As stated previously, you can run ColdFusion for free on your development server with no timeout restrictions.

What jQuery Mobile does with forms

Before we get into the code, there are two very important things you should know about what jQuery Mobile will do with your HTML forms:

- All forms will submit their data via AJAX. That means the data is sent directly to the action of your form and the result will be brought back to the user and placed within the page that held the form. This prevents a full page reload.

- All form fields are automatically enhanced, each in their own way. As we go on in the chapter, you will see examples of this; but basically jQuery Mobile modifies your form fields to work better on a mobile device. A great example of this is the buttons. jQuery Mobile automatically widens and heightens buttons to make them easier to click in the small form factor of a phone. If for some reason you don't like this, jQuery Mobile provides a way to disable this, either on a global or per use basis.

With that in mind, let's look at our first example in code 6-1:

```
code 6-1: test1.html
<div data-role="header">
  <h1>Form Demo</h1>
</div>
<div data-role="content">
  <form action="echo.cfm" method="post">
    <div data-role="fieldcontain">
      <label for="name">Name:</label>
      <input type="text" name="name" id="name" value="" />
    </div>
    <div data-role="fieldcontain">
      <label for="email">Email:</label>
      <input type="text" name="email" id="email" value="" />
    </div>
    <div data-role="fieldcontain">
      <input type="submit" name="submit" value="Send" />
    </div>
  </form>
</div>
```

As before, we've focused on the important part of the template, the content within our main `page` `div` block. Notice that, for the most part, this is a generic form, but that every label and field is wrapped with the following `div`:

```
<div data-role="fieldcontain">
</div>
```

This will help jQuery Mobile align the label and form field. You'll see why in a moment. Our form has two text fields, one for name and one for e-mail. The last item is just the submit button. So, outside of using a `fieldcontain` wrapper and ensuring we have labels for our form fields, nothing special is going on here. Right away, though, you can see some pretty impressive changes to the form:

Notice how the labels are presented above the form fields. This gives the fields more space on the mobile device. Also, notice the submit button is large and easy to click. If we rotate the device, jQuery Mobile updates the display to take advantage of the additional space:

Notice that the fields now line up directly to the right of their labels. So what happens when the form is submitted? As mentioned at the beginning of this chapter, we're making use of ColdFusion to handle responding to the form requests. Our `echo.cfm` template will simply loop over all the form fields and display them back out to the user:

```
code 6-2: echo.cfm
<div data-role="page">
  <div data-role="header">
    <h1>Form Result</h1>
  </div>
  <div data-role="content">
    <cfloop item="field" collection="#form#">
      <cfoutput>
        <p>
          The form field #field# has the value #form[field]#.
        </p>
      </cfoutput>
    </cfloop>
  </div>
</div>
```

If you do not want to install ColdFusion, you can simply edit the form action value in `code 6-1` to point to a PHP file, or any other server-side processor. You may also simply change it to `test1.html`, the file itself. Nothing will happen when you submit it, but you will not get an error either. Here's what the device will show after hitting submit:

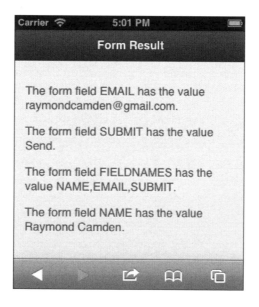

Another great example of how jQuery Mobile updates form fields is with `textarea`. `textarea`, by default, can be very difficult to work with on a mobile device, especially as the amount of text grows beyond the size of the `textarea` and a scroll bar is added. In the following code snippet, we've simply modified the previous form to include a third item, a bio field that uses `textarea`. The complete file may be found in the book's code ZIP file as `test2.html`. The following code snippet is the `div` block added after the previous two fields:

```
<div data-role="fieldcontain">
  <label for="bio">Bio:</label>
  <textarea name="bio" id="bio"></textarea>
</div>
```

When viewed on the device, the `textarea` expands to take in more width like the regular text fields, and grows taller.

But once you begin typing and entering multiple lines of text, the `textarea` automatically expands:

This is much easier to read than if scrollbars had been used. Now let's look at another common form option: radio buttons and checkboxes.

Working with radio buttons and checkboxes

Both radio buttons and checkboxes are also updated to work nicely on a mobile device, but require a tiny bit more code. In the earlier examples, we wrapped form fields with a `div` tag that made use of `data-role="fieldcontain"`. When working with radio buttons and checkboxes, one more tag is required.

```
<fieldset data-role="controlgroup">
```

This `fieldset` tag will be used to group together your radio buttons or checkboxes. `code 6-3` demonstrates one set of radio buttons and one checkbox group:

```
code 6-3: test3.html
<div data-role="page">
  <div data-role="header">
    <h1>Form Demo</h1>
  </div>
```

```
<div data-role="content">
  <form action="echo.cfm" method="post">
    <div data-role="fieldcontain">
      <fieldset data-role="controlgroup">
        <legend>Favorite Movie:</legend>
        <input type="radio" name="favoritemovie"
          id="favoritemovie1" value="Star Wars">
        <label for="favoritemovie1">Star Wars</label>
        <input type="radio" name="favoritemovie"
          id="favoritemovie2" value="Vanilla Sky">
        <label for="favoritemovie2">Vanilla Sky</label>
        <input type="radio" name="favoritemovie"
          id="favoritemovie3" value="Inception">
        label for="favoritemovie3">Inception</label>
      </fieldset>
    </div>
    <div data-role="fieldcontain">
      <fieldset data-role="controlgroup">
        <legend>Favorite Colors:</legend>
        <input type="checkbox" name="favoritecolor"
          id="favoritecolor1" value="Green">
        <label for="favoritecolor1">Green</label>
        <input type="checkbox" name="favoritecolor"
          id="favoritecolor2" value="Red">
        <label for="favoritecolor2">Red</label>
        <input type="checkbox" name="favoritecolor"
          id="favoritecolor3" value="Yellow">
        <label for="favoritecolor3">Yellow</label>
      </fieldset>
    </div>
    <input type="submit" name="submit" value="Send"  />
    </div>
  </form>
</div>
</div>
```

Our form has two main questions: what is your favorite movie and what are your favorite colors? Each block is wrapped in the `div` tag we mentioned before. Inside this is the `fieldset` using `data-role="controlgroup"`. Finally, you then have your radio and checkbox groups. It is important to include the labels with each input control, as each of the previous examples have. Once rendered, jQuery Mobile groups these into a nice-looking, singular control.

Notice the wide, clickable regions for each item. This makes it much easier to select items on a mobile device. Another interesting feature of both of these controls is the ability to turn them into horizontal button bars. In `test4.html`, both `fieldset` tags were modified to include a new data attribute:

```
<fieldset data-role="controlgroup" data-type="horizontal">
```

As you can see, the effect doesn't work well with the longer text in the first group, so be sure to test it.

Working with select menus

Yet another example of jQuery Mobile form enhancement is with `select` menus. As with our earlier examples, we make use of a `fieldcontain` div and `label` tag, but outside of that, the `select` tag is used as normal. The following code snippet is from `test5.html`:

```
<div data-role="fieldcontain">
  <label for="favmovie">Favorite Movie:</label>
  <select name="favmovie" id="favmovie">
    <option value="Star Wars">Star Wars</option>
    <option value="Revenge of the Sith">Revenge of the
      </option>
    <option value="Tron">Tron</option>
    <option value="Tron Legacy">Tron Legacy</option>
  </select>
</div>
```

On the mobile device, the initial display of the select control is modified to be easier to hit.

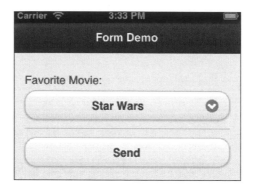

However, once clicked, the device's native menu will take over. This will look different on the platform you are using. The following screenshot shows how iOS renders the menu:

The following screenshot demonstrates how Android renders it:

Another option to use with the select fields is grouping. jQuery Mobile allows you to vertically or horizontally group together multiple select fields. In both cases, all that's required is to wrap your select fields in a fieldset using the data-role of controlgroup, much like we did earlier for radio and checkboxes. The following code snippet is an example of a vertically aligned group of select fields:

```
<div data-role="fieldcontain">
  <fieldset data-role="controlgroup">
    <legend>Trip Setup:</legend>
    <label for="location">Location</label>
    <select name="location" id="location">
      <option value="Home">Home</option>
      <option value="Work">Work</option>
      <option value="Moon">Moon</option>
      <option value="Airport">Airport</option>
    </select>
    <label for="time">Time</label>
    <select name="time" id="time">
      <option value="Morning">Morning</option>
      <option value="Afternoon">Afternoon</option>
      <option value="Evening">Evening</option>
    </select>
    <label for="time">Meal</label>
    <select name="meal" id="meal">
      <option value="Meat">Meat</option>
      <option value="Vegan">Vegan</option>
```

```
      <option value="Kosher">Kosher</option>
    </select>
  </fieldset>
</div>
```

The rest of this template can be found in `test6.html`. The following screenshot shows how it looks:

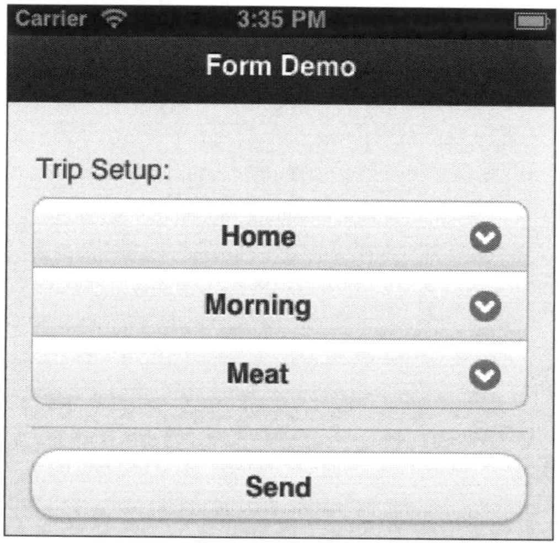

Note how jQuery Mobile groups them together and nicely rounds the corners. The horizontal version can be achieved by adding a `data-type="horizontal"` attribute to the `fieldset` tag. It's also important to remove the `div` using `"fieldcontain"`. Here is an example (the complete file can be found in `test7.html`):

```
<form action="echo.cfm" method="post">
  <fieldset data-role="controlgroup" data-type="horizontal">
    <legend>Trip Setup:</legend>
    <label for="location">Location</label>
    <select name="location" id="location">
      <option value="Home">Home</option>
      <option value="Work">Work</option>
      <option value="Moon">Moon</option>
      <option value="Airport">Airport</option>
    </select>
    <label for="time">Time</label>
    <select name="time" id="time">
      <option value="Morning">Morning</option>
      <option value="Afternoon">Afternoon</option>
      <option value="Evening">Evening</option>
    </select>
```

```
    <label for="meal">Meal</label>
    <select name="meal" id="meal">
      <option value="Meat">Meat</option>
      <option value="Vegan">Vegan</option>
      <option value="Kosher">Kosher</option>
    </select>
  </fieldset>
  <div data-role="fieldcontain">
    <input type="submit" name="submit" value="Send" />
  </div>
</form>
```

The following screenshot shows the result:

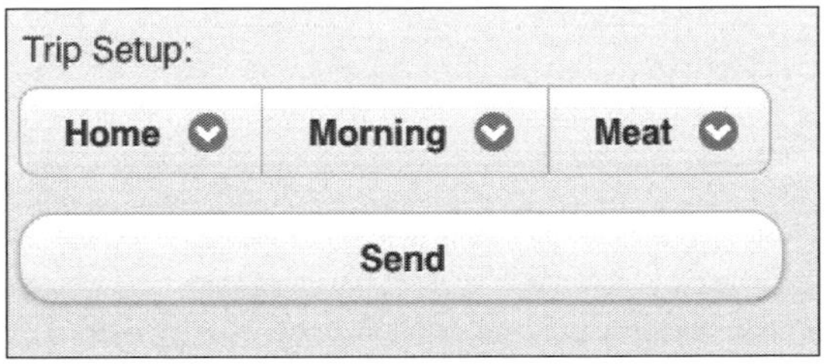

Search, toggle, and slider fields

Along with taking regular form fields and making them work better, jQuery Mobile also helps make some of the newer HTML5 form fields work correctly across multiple browsers. While support still isn't nailed down on the desktop across every major browser, jQuery Mobile provides built-in support for search, toggle, and slider fields. Let's take a look at each one.

Search fields

The simplest of the three new fields, search fields simply adds a quick delete icon to the end of the field after you begin typing. Some devices will also put an hourglass icon in front to help convey the idea of the field being used for some type of search. To use this field, simply switch your type from text to search, as in the following example from test8.html:

```
<div data-role="fieldcontain">
  <label for="name">Name:</label>
  <input type="search" name="name" id="name" value=""  />
</div>
```

The following screenshot is the result. Notice that I've typed a bit and the field automatically adds a **Delete** icon at the end.

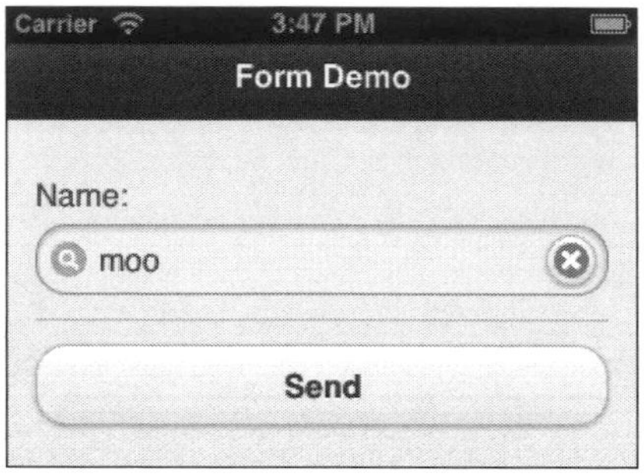

Flip toggle fields

Flip toggle fields are controls that flip back between one and two values. Creating a toggle field involves using a select control with a particular data-role value. Now, here is where things may get a bit confusing. To enable a `select` field to turn into a toggle field, you use `data-role="slider"`. In a moment, we're going to look at another slider control, but it uses a different technique. Just keep in mind that even though you'll be seeing `"slider"` in the HTML, it's really a toggle control we are creating. Let's look at a simple example. (You can find the complete source for this in `test9.html`.)

```
<div data-role="fieldcontain">
  <label for="gender">Gender:</label>
  <select name="gender" id="gender" data-role="slider">
    <option value="0">Male</option>
    <option value="1">Female</option>
  </select>
</div>
```

The following screenshot demonstrates how the flip toggle is displayed:

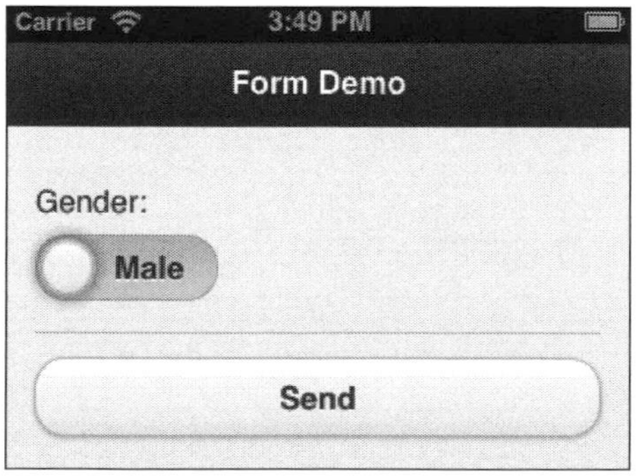

Slider fields

For the last of our special fields, we take a look at sliders. Like search fields, this is based on an HTML5 specification that works in some browsers and not others. jQuery Mobile simply makes it work everywhere. To enable this field, we take a regular text field and switch the type to `"range"` To give our slider a range, we also provide a `min` and `max` value. You can also add additional color to the slider by adding the attribute `data-highlight="true"`. The following code snippet is a sample. (You can find the complete file in `test10.html`.)

```
<div data-role="fieldcontain">
  <label for="coolness">Coolness:</label>
  <input type="range" name="coolness" id="coolness" min="0"
    max="100"
    value="22" data-highlight="true">
</div>
```

The result is a slider control and an input field. Both allow you to modify the value between the minimum and maximum value.

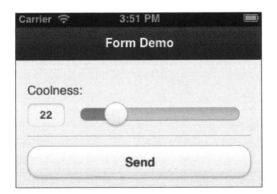

> The HTML5 specification for the range supports a step attribute. While this works in some browsers, it is not yet directly supported by jQuery Mobile.

In other words, jQuery Mobile won't try to add this support on a browser that doesn't have it built-in. You can add the attribute as long as you are aware it may not always work as intended.

You can also get more fancy with your slider control by creating a dual range slider. In this case, your slider has two controls, allowing your user to select both a minimum and maximum value. To create this control, you first create a div block with the data-role of range slider:

```
<div data-role="rangeslider">
</div>
```

Then within the div attribute, you simply add two slider controls, using the exact same syntax you used before. In the following code snippet (taken from test11. html), you can see an example of this:

```
<div data-role="fieldcontain">
  <div data-role="rangeslider">
    <label for="coolnessLow">Cool Range:</label>
    <input type="range" name="coolnessLow" id="coolnessLow"
      min="0" max="100" value="22">
    <input type="range" name="coolnessHigh" id="coolnessHigh"
      min="0" max="100" value="82">
  </div>
</div>
```

A few things to note here; there is only one label. If you provide two labels, the second one is automatically hidden. Next, in theory, even though both ranges have the same minimum and maximum, jQuery Mobile will not let you drag the second value below the first. The following is a screenshot of this in action:

Using native form controls

Now you've seen how far jQuery Mobile will go to enhance and empower your form fields to work better on mobile devices. But what if you don't like what jQuery Mobile does? What if you love its updates to buttons but despise its changes to drop-downs? Luckily jQuery Mobile provides a simple way to disable automatic enhancement. In each field you want to be left alone, simply add `data-role="none"` to the markup. So given the following HTML, the first item will be updated while the second will not:

```
<input type="submit" value="Awesome">
<input type="submit" value="Not So Awesome" data-role="none">
```

You can see an example of this in `test12.html`.

Another option is to disable it when jQuery Mobile is initialized. That option will be discussed in *Chapter 9, jQuery Mobile Configuration, Utilities, and JavaScript Methods*.

Working with the mini fields

In the previous examples, we saw how jQuery Mobile automatically enhances form fields to make them easier to use on smaller, touch based devices. In general, jQuery Mobile took your fields and made them nice and fat. While that's desirable most of the time, you may want to put your form fields on a bit of a diet. This is especially true for placing form fields in a header or footer. jQuery Mobile supports an attribute on any form field that creates a smaller version of the field: `data-mini="true"`. The following code snippet is a complete example:

```
<div data-role="fieldcontain">
  <label for="name">Name:</label>
  <input type="search" name="name" id="name" value=""  />
</div>
<div data-role="fieldcontain">
  <label for="name">Name (Slim):</label>
  <input type="search" name="name" id="name" value="" data-
    mini="true"  />
</div>
```

The result is a bit subtle, but you can see the height difference in the second field in the following screenshot:

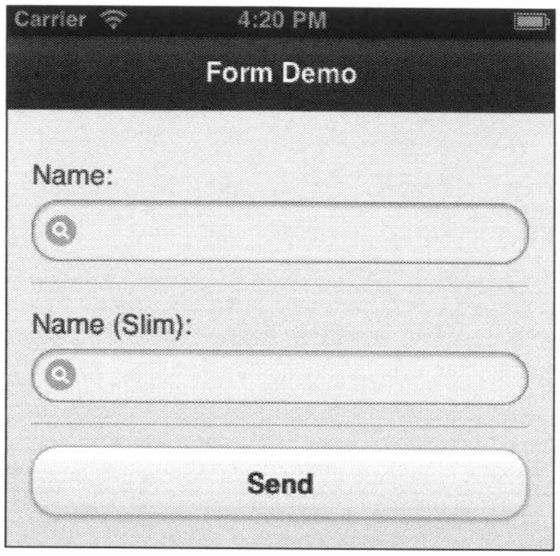

This example may be found with the rest of the files in a file named `test13.html`.

Summary

In this chapter, we discussed forms and how they are rendered in a jQuery Mobile application. We discussed how jQuery Mobile automatically turns all form submissions into AJAX-based calls and updates form fields to work better on mobile devices. Not only are all your form fields automatically updated, but you can also make use of new controls like the toggle, slider, and search inputs.

In the next chapter, we'll take a look at modal dialogs, widgets, and layout grids. These provide additional UI options for your mobile optimized site.

7
Creating Modal Dialogs and Widgets

In this chapter, we will look at dialogs, grids, and other widgets. In the previous chapters we've dealt with pages, buttons, and form controls. While jQuery Mobile provides great support for them, you get even more UI controls than you get within the framework.

In this chapter, we will do the following:

- Discuss how to link to and create dialogs; also how to handle leaving them
- Demonstrate grids and how you can add them to your pages
- Show how collapsible blocks allow you to pack a lot of information in a small amount of space
- Explain how to create popups and how they differ from dialogs
- Discuss the new responsive table and panel widgets

Creating dialogs

Dialogs, at least under the jQuery Mobile framework, are small windows that cover an existing page. They typically provide a short message or question for the user. They will also typically include a button that allows the user to dismiss the dialog and return back to the site. Creating a dialog in jQuery Mobile is done by adding a simple attribute to a link: `data-rel="dialog"`. The following code snippet demonstrates an example:

```
Code 7-1: test1.html
<div data-role="page" id="first">
    <div data-role="header">
      <h1>Dialog Test</h1>
```

```
    </div>
    <div data-role="content">
      <p>
        <a href="#page2">Another Page (normal)</a>
      </p>
      <p>
        <a href="#page3" data-rel="dialog">A Dialog (dialog)</a>
      </p>
      </div>
    </div>
    <div data-role="page" id="page2">
      <div data-role="header">
        <h1>The Second</h1>
      </div>
      <div data-role="content">
        <p>
          This is the Second
        </p>
      </div>
    </div>
    <div data-role="page" id="page3">
      <div data-role="header">
        <h1>The Third</h1>
      </div>
      <div data-role="content">
      <p>
        This is the Third
      </p>
    </div>
  </div>
```

This is a simple, multi-page jQuery Mobile site. Notice how we link to the second
and third page. The first link is typical. The second link, though, includes the
data-rel attribute mentioned earlier. Notice that both the second and third pages
are defined in the usual manner. So, the only change we have here is in the link.
When that second link is clicked, the page is rendered completely differently,
as shown in the following screenshot:

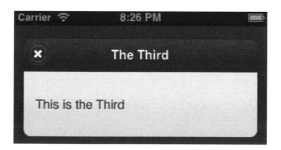

Remember that the page wasn't defined differently. The change you see in the preceding screenshot is driven by the change to the link itself. That's it! Clicking the little **X** button will hide the dialog and return the user back to the original page.

Any link within the page will handle closing the dialog as well. If you wish to add a cancel type button, or link, you can do so using `data-rel="back"` in the link. The target of the link should be to the page that launched the dialog. `Code 7-2` shows a modified version of the earlier template. In this one, we've simply added two buttons to the dialog. The first button will launch the second page, while the second one will act as a cancel action:

```
Code 7-2: test2.html
<div data-role="page" id="first">
    <div data-role="header">
      <h1>Dialog Test</h1>
    </div>
    <div data-role="content">
      <p>
        <a href="#page2">Another Page (normal)</a>
      </p>
      <p>
        <a href="#page3" data-rel="dialog">A Dialog (dialog)</a>
      </p>
    </div>
  </div>
  <div data-role="page" id="page2">
    <div data-role="header">
      <h1>The Second</h1>
    </div>
  <div data-role="content">
    <p>
      This is the Second
    </p>
  </div>
</div>
<div data-role="page" id="page3">
  <div data-role="header">
    <h1>The Third</h1>
  </div>
    <div data-role="content">
      <p>
        This is the Third
      </p>
        <a href="#page2" data-role="button">Page 2</a>
```

```
      <a href="#first" data-role="button" data-
        rel="back">Cancel</a>
    </div>
  </div>
```

The major change in this template is the addition of the buttons in the dialog, contained within `page3` div. Notice that the first link is turned into a button, but outside of that is a simple link. The second button includes the addition of the `data-rel="back"` attribute. This will handle simply dismissing the dialog. The following screenshot shows how the dialog looks with the buttons added:

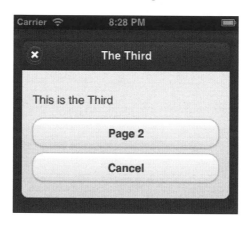

Laying out content with grids

Grids are one of the few features of jQuery Mobile that do not make use of particular data attributes. Instead, you work with grids simply by specifying CSS classes for your content.

Grids come in four flavors: two-column, three-column, four-column, and five-column. (You will probably not want to use the five-column grid on a mobile phone device. Save that for a tablet instead.)

You begin a grid with a `div` block that makes use of the `ui-grid-X` class, where `X` will be either `a`, `b`, `c`, or `d`. `ui-grid-a` represents a two-column grid, whereas `ui-grid-b` is a three-column grid. You can probably guess what `c` and `d` create.

So to begin a two-column grid, you would wrap your content with the following:

```
<div class="ui-grid-a">
  Content
</div>
```

Within the `div` tag, you then use a `div` for each "cell" of the content. The class for grid calls begin with `ui-block-X`, where `X` goes from `a` to `d`. `ui-block-a` would be used for the first cell, `ui-block-b` for the next, and so on. This works much like HTML tables.

Putting it together, the following code snippet demonstrates a simple two-column grid with two cells of content:

```
<div class="ui-grid-a">
  <div class="ui-block-a">Left</div>
  <div class="ui-block-b">Right</div>
</div>
```

Text within a cell will automatically get wrapped. `Code 7-3` demonstrates a simple grid with a large amount of text in one of the columns:

```
Code 7-3: test3.html
<div data-role="page" id="first">
    <div data-role="header">
      <h1>Grid Test</h1>
    </div>
    <div data-role="content">
      <div class="ui-grid-a">
        <div class="ui-block-a">
        <p>
        This is my left hand content. There won't be a lot of
        it.
        </p>
        </div>
        <div class="ui-block-b">
          <p>
            This is my right hand content. I'm going to fill it
            with some dummy text.
          </p>
          <p>
            Bacon ipsum dolor sit amet andouille capicola spare
            ribs, short loin venison sausage prosciutto
            turkey flank frankfurter pork belly short ribs.
            chop, pancetta turkey bacon short ribs ham flank
            pork belly. Tongue strip steak short ribs tail
          </p>
        </div>
      </div>
    </div>
</div>
```

In the mobile browser, you can clearly see the two columns, as shown in the following screenshot:

If the text in those div tags seems a bit close together, there is a simple fix for that. In order to add a bit more space between the content of the grid cells, you can add a class to your main div that specifies ui-content. This tells jQuery Mobile to pad the content a bit. So, for example, consider the following line of code:

```
<div class="ui-grid-a ui-content">
```

This small change modifies the previous screenshot in the following manner:

Working with other types of grids then is simply a matter of switching to the other classes. For example, a four-column grid would be set up similar to the following code snippet:

```
<div class="ui-grid-c">
  <div class="ui-block-a">1st cell</div>
  <div class="ui-block-b">2nd cell</div>
  <div class="ui-block-c">3rd cell</div>
</div>
```

Again, keep in mind your target audience. Anything over two columns may be too thin on a mobile phone.

To create multiple rows in a grid, you simply need to repeat the blocks. The following code snippet demonstrates a simple example of a grid with two rows of cells:

```
<div class="ui-grid-a">
  <div class="ui-block-a">Left Top</div>
  <div class="ui-block-b">Right Top</div>
  <div class="ui-block-a">Left Bottom</div>
  <div class="ui-block-b">Right Bottom</div>
</div>
```

Notice that there isn't any concept of a row. jQuery Mobile handles knowing that it should create a new row when the block starts over with the one marked ui-block-a. The following code snippet, Code 7-4, is a simple example:

```
Code 7-4:test4.html
<div data-role="page" id="first">
    <div data-role="header">
      <h1>Grid Test</h1>
    </div>
    <div data-role="content">
      <div class="ui-grid-a">
        <div class="ui-block-a">
          <p>
            <img src="ray.png">
          </p>
        </div>
        <div class="ui-block-b">
        <p>
        This is Raymond Camden. Here is some text about him. It
        may wrap or it may not but jQuery Mobile will make it
        look good. Unlike Ray!
        </p>
        </div>
        <div class="ui-block-a">
```

```
    <p>
       This is Scott Stroz. Scott Stroz is a guy who plays
       golf and is really good at FPS video games.
    </p>
  </div>
  <div class="ui-block-b">
    <p>
       <img src="scott.png">
    </p>
  </div>
 </div>
</div>
</div>
```

The following screenshot shows the result:

Making responsive grids

Earlier in the chapter, we mentioned that complex grids may not work depending on the size of your targeted devices. A simple two-column grid is fine, but larger grids would render well only on tablets. Luckily, there's a simple solution for it. jQuery Mobile's latest updates include a much better support for responsive design. Let's consider a simple example. The following is a screenshot of a web page using a four-column grid:

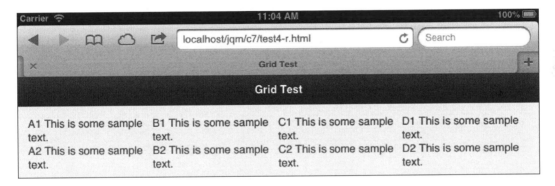

It is surely readable, but a bit dense. The same page in a table, though, will be more readable. The following is a screenshot of the same page on an iPad:

By making use of responsive design, we can handle the different sizes intelligently using the same basic HTML. jQuery Mobile enables a simple solution for this by adding the `ui-responsive` class to an existing grid. The following is an example:

```
<div class="ui-grid-c ui-responsive">
```

By making this one small change, look how the phone version of our page changes:

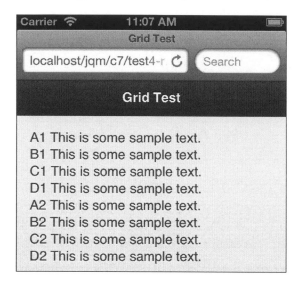

The four-column layout now is a one-column layout instead. If viewed in a tablet, the original four-column design is preserved.

Working with collapsible content

The next widget we will look at in this chapter supports collapsible content. This is simply content that can be collapsed and expanded. Creating a collapsible content widget is as simple as wrapping it in a `div`, adding `data-role="collapsible"`, and including a title for the content. Consider the following simple example:

```
<div data-role="collapsible">
  <h1>My News</h1>
  <p>This is the latest news about me…
</div>
```

Upon rendering, jQuery Mobile will turn the title into a clickable banner that can expand and collapse the content within. Let's look at a real example. Imagine you want to share the location of your company's primary address. You also want to include satellite offices. Because most people won't care about the other offices, we can use a simple collapsible content widget to hide the content by default. The following code snippet, `Code 7-5`, demonstrates an example of this:

```
Code 7-5: test5.html
<div data-role="page" id="first">
    <div data-role="header">
```

```
      <h1>Our Offices</h1>
  </div>
  <div data-role="content">
    <p>
      <strong>Main Office:</strong><br/>
      400 Elm Street<br/>
      New York, NY<br/>
      90210
    </p>
    <div data-role="collapsible">
      <h3>Satellite Offices</h3>
        <p>
          <strong>Asia:</strong>
          Another Address Here
        </p>
        <p>
          <strong>Europe:</strong>
          Another Address Here
        </p>
        <p>
          <strong>Mars:</strong>
          Another Address Here
        </p>
    </div>
  </div>
</div>
```

You can see that the other offices are all wrapped in the div tag using the new collapsible content role. When viewed, they are seen to be hidden:

Clicking on the + icon next to the title opens it, and can be clicked again to close it:

By default, jQuery Mobile will collapse and hide the content. You can, of course, tell jQuery Mobile to initialize the block as open instead of closed. To do so, simply add `data-collapsed="false"` to the initial `div` tag. For example, consider the following code snippet:

```
<div data-role="collapsible" data-collapsed="false">
  <h1>My News</h1>
  <p>This is the latest news about me…
</div>
```

This region will still have the ability to collapse and open, but will simply default to being opened initially.

Another option for collapsible content blocks is the ability to theme the content of the area that is collapsed. By providing a `data-content-theme` attribute, you can specify a background color that makes the region a bit more cohesive. Theming is covered in *Chapter 11, Enhancing jQuery Mobile* but we can take a look at a quick example. In the following screenshot, the first region does not make use of the feature, while the second one does:

Notice that the icon is also shifted to the right. This demonstrates another option: `data-iconpos`. The following code snippet, found in the `test5-2.html` file in the code folder, demonstrates these options:

```
<div data-role="collapsible">
  <h3>First</h3>
  <p>
    Hello World...
  </p>
</div>
<div data-role="collapsible" data-content-theme="c" data-
  iconpos="right">
  <h3>First</h3>
  <p>
    Hello World again...
  </p>
</div>
```

Finally, you can take multiple collapsible regions and combine them into one called an **accordion**. This is done by simply taking multiple collapsible blocks and wrapping them in a new div tag. This div tag makes use of data-role="collapsible-set" to make the inner blocks as one unit. Code 7-6 demonstrates an example of this. It takes the earlier office address example and uses a collapsible set for each unique address:

```
Code 7-6: test6.html
<div data-role="page" id="first">
    <div data-role="header">
      <h1>Our Offices</h1>
    </div>
    <div data-role="content">
      <div data-role="collapsible-set">
        <div data-role="collapsible">
          <h3>Main Office</h3>
            <p>
              400 Elm Street<br/>
              New York, NY<br/>
              90210
            </p>
        </div>
        <div data-role="collapsible">
          <h3>Asia</h3>
            <p>
              Another Address Here
            </p>
        </div>
        <div data-role="collapsible">
          <h3>Europe</h3>
            <p>
              Another Address Here
            </p>
        </div>
        <div data-role="collapsible">
          <h3>Mars</h3>
            <p>
              Another Address Here
            </p>
        </div>
      </div>
    </div>
</div>
```

In `Code 7-6`, we simply wrap four collapsible blocks with a `div` tag that makes use of a collapsible set. Once done, jQuery Mobile will group them together and automatically close one once another is open:

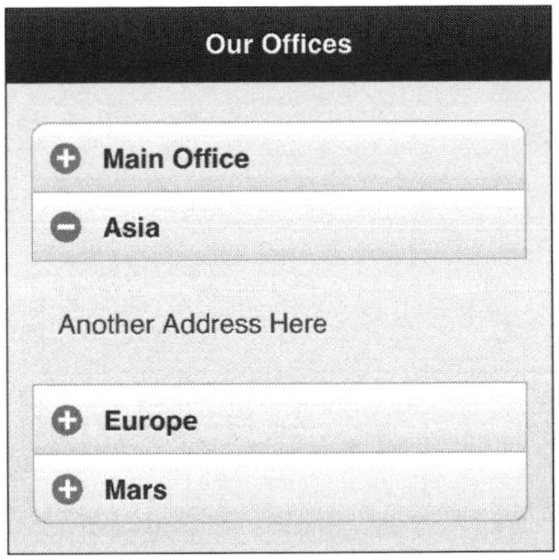

Popups

Popups (also known as tooltips) are similar to dialogs, but are much smaller and not necessarily model. They can be useful for providing contextual help or descriptive text. Like dialogs, you create a popup by applying a data attribute to a link. Unlike dialogs, your "target" for the popup content will be another `div` in the same page. Let's look at an example:

```
Code 7-7: test7.html
<div data-role="page" id="first">

    <div data-role="header">
       <h1>Popup Demo</h1>
    </div>

    <div data-role="content">

        <a href="#firstPopup" data-role="button" data-rel="popup">Show
Popup</a>

    </div>

    <div data-role="popup" id="firstPopup">
        <p>This is just a test. It has some text in it. It is incredibly
awesome.</p>
    </div>

</div>
```

Inside our main content `div` is a simple link. In order to let jQuery Mobile know this is linking to a popup, the `data-rel` attribute is specified. Note that the `ID` value points to another `div` on the page. That `div` has the role of a popup. When viewed in the browser, jQuery Mobile automatically hides this `div`. It will only show up when the link is actually clicked:

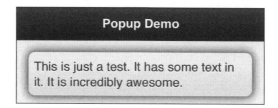

As with other widgets, you have multiple options you can specify for your popups. They include position options, modal properties, transitions (for opening and closing), and of course, theming. Let's look at the position and modal properties first, as they are the most interesting.

By default, popups will be positioned to the center of the item that launched them. So for example, in our previous code this was a button. You can modify this value by using a `data-position-to` attribute. The default is `origin`. You can center the popup to the entire window by using, you guessed it, `window`. Finally, you can also pass a jQuery selector as a value and that DOM item will be used to center the popup.

Normally, popups are dismissed if you click anywhere on the page. You can create a modal popup (which acts just like the dialogs discussed earlier in the chapter) by setting them as non-dismissible. This is done by adding `data-dismissible="false"` to the `div` containing the popup. To be clear, this is done to the `div`, not the link. Code 7-8 demonstrates an example of both of these options:

```
Code 7-8: test8.html
<div data-role="page" id="first">

    <div data-role="header">
       <h1>Advanced Popup Demo</h1>
    </div>

    <div data-role="content">

        <a href="#firstPopup" data-role="button"
        data-rel="popup" data-position-to="window">Show Popup</a>
        <p>
            I'm including some text here just so that we can
            properly demo the new position of my popup. It
```

```
        will be centered on the window, not the link
        above. That is <i>awesome</i>.
    </p>
  </div>

  <div data-role="popup" id="firstPopup" data-dismissible="false"
class="ui-content">
    <p>This is just a test. It has some text in it. It is incredibly
awesome.</p>
    <a href="" data-rel="back" data-theme="b" data-role="button">Get
Rid Of Me</a>
  </div>

</div>
```

This template is mostly like the previous one, but with a few important updates. Firstly, make note of the `data-position-to` attribute. This will position the popup centered to the window. We added some text to help make the page a bit taller in general. The next change was to the popup `div`. First, we added the `data-dismissible="false"` attribute to make it more of a modal dialog. In order to actually get rid of the popup though we have to add our own UI. To do this, a button was added. By specifying `data-rel="back"`, jQuery Mobile will handle getting rid of the popup.

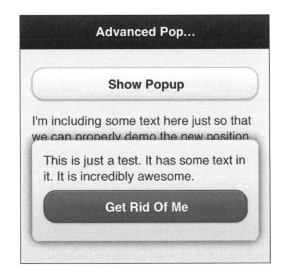

Responsive tables

Tables present a particular difficulty for mobile browsers. Generally, tables consist of a large amount of data. This could take up an entire screen of a desktop browser. On a mobile browser, this can be even more condensed. In the folder for this chapter's source code, take a look at `test_table.html`. We won't print the code here, but it is a rather simple four-column, four-row table. On a mobile device, this information fits, but just barely.

Table Demo			
Name	**Title**	**Age**	**Beers**
Raymond Camden	Jedi Master	40	3
Scott Stroz	Golf Master	45	9
Todd Sharp	Football Hero	36	5
Dave Ferguson	Ninja Star	32	8

jQuery Mobile can make this work better by creating a responsive version of the table. There are a few small changes you must make to your tables to enable this feature. Firstly, ensure your table makes use of `thead` and `tbody` blocks. The previous example, `test_table.html`, did this already. Code 7-9 demonstrates what else we have to do to make the table responsive:

```
Code 7-9: test_resp_table.html
<table data-role="table" class="ui-responsive table-stroke">
   <thead>
      <tr>
         <th>Name</th>
         <th>Title</th>
         <th>Age</th>
         <th>Beers</th>
      </tr>
   </thead>
   <tbody>
      <tr>
         <th>Raymond Camden</th>
         <td>Jedi Master</td>
         <td>40</td>
         <td>3</td>
      </tr>
      <tr>
```

```
            <th>Scott Stroz</th>
            <td>Golf Master</td>
            <td>45</td>
            <td>9</td>
        </tr>
    </tbody>
</table>
```

Firstly, we add a `data-role="table"` to the core table block. This will enable basic responsive functionality for the table. To use the default responsive breakpoints, a class of `ui-responsive` is also added. There is one more small change we will explain in a minute. Now take a look at the table in the following screenshot:

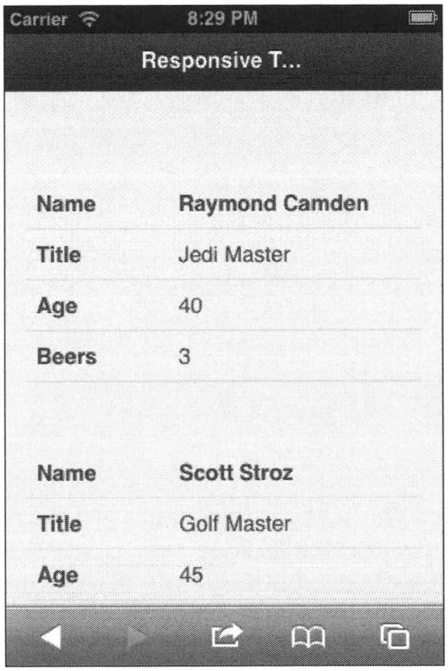

jQuery Mobile has broken the table up into blocks of content that are *much* easier to read. Even better, if you view this table on a larger screen tablet, it will render as a regular table. Remember the one other change we mentioned? Note the vertical space between the first "row" (for **Raymond Camden**) and the second "row" (**Scott Stroz**). By default, jQuery Mobile will not add the vertical spacing. This was accomplished by changing the first cell of each row to a `<th>` element. These elements are normally reserved for the top-level header, but can be used inside a `tbody` as well. For our demo here, they serve to help visually group the different rows in a more visually pleasing manner.

But wait, there's more! jQuery Mobile also supports another powerful feature—the ability to selectively filter the columns displayed to the user. The **Column Toggle** feature is enabled by adding `data-mode="columntoggle"` to a table. As an example:

```
<table data-role="table" data-mode="columntoggle" class="ui-responsive
table-stroke" id="peopleTable">
```

This adds a column toggle widget to your table. (Note that, in order for the button to associate with your table, you must include an ID). By default, though, the column toggle widget will not actually allow you to select any columns. To provide a list of columns, selectively add a data-priority to each row header that you wish to make toggle-enabled. You *do not* need to select every column, nor do you need to provide any specific order. The following is an example taken from the `test_resp_adv_ table.html` file:

```
<table data-role="table" data-mode="columntoggle" class="ui-responsive
table-stroke" id="peopleTable">
    <thead>
        <tr>
            <th>Name</th>
            <th data-priority="1">Title</th>
            <th data-priority="2">Age</th>
            <th data-priority="3">Beers</th>
        </tr>
    </thead>
```

Again, the preceding order is totally arbitrary. While the order will probably match the order of your columns, you are free to choose any order you wish. Now, when the table is displayed, a column toggle widget is added and the default display is focused on the first priority column:

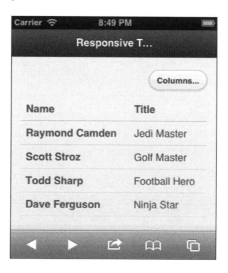

Clicking on the widget opens a popup:

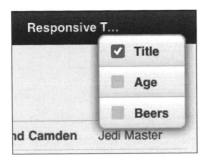

The user can then select which columns they care about, as illustrated in the following screenshot:

And that's it! In case you're curious, you can change the text on the column toggle widget by simply using the `data-column-btn-text` attribute on the table.

Working with panels

The last widget we will discuss is the Panel widget. The Panel widget sits on the left or right-hand side of your page and can be shown or hidden dynamically. Because of the limited "real estate" on mobile devices, panels are a nice way to hide navigation or other content until the user requests it.

Creating a panel in jQuery Mobile is simple. First, add a new `<div>` block with a `data-role` of panel. This must be done inside the `div` that defines your page, and should be outside of the content region. Code 7-10 demonstrates an example of this.

```
Code 7-10: test_panel.html
<div data-role="page" id="first">

    <div data-role="panel" id="leftPanel">
        This is the left panel.

        <p>
            <a data-role="button" data-rel="close">Close</a>
        </p>

    </div>

    <div data-role="header">
        <h1>Test Panel</h1>
    </div>

    <div data-role="content">

        <a href="#leftPanel" data-role="button">Open My Panel!</a>
        <a href="#rightPanel" data-role="button">Open Another Panel!</a>

    </div>


    <div data-role="panel" data-position="right" id="rightPanel">
        This is the right panel.

        <p>
            <a data-role="button" data-rel="close">Close</a>
        </p>

    </div>

</div>
```

In this example, two panels are defined, one above and one below the content div. Note that both have an ID. This is required so that the panel can be opened. Opening panels is done with a simple link. Inside the content div are two links—one for the left-hand side panel and one for the right-hand side panel. Panels are placed on the left-hand side by default. To place one on the right-hand side, add the data-position attribute. The final aspect to this is how panels are closed. By default, a panel will be closed if the user clicks anywhere *outside* the panel. Panels will also be closed if the user swipes. But because a user may not know this, it makes sense to also provide a manual way to close the panel. In both panels, a link with a data-rel value of close is provided. The following is an example with the left-hand side panel opened:

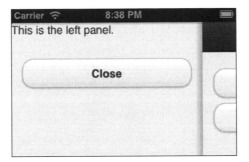

Notice that the main page content was pushed to the right-hand side. Panels are displayed in one of three ways and can be customized via the data-display attribute. The default value is reveal which pushes the content away. Another option is overlay. As you can guess, this will render the panel on top of the main page content, as shown in the following screenshot:

The final option is push. The push and reveal options look very similar, and frankly, many users may not be able to tell the difference. The reveal option will move the main page content to the side *revealing* the panel. The push option acts as if the panel is *pushing* the content to the side. You can see an example of both of these in test_panel_display.html.

Finally, you can control the close behavior of panels by using `data-swipe-close="false"` and `data-dismissable="false"`. The former disables the swipe-to-close behavior and the latter prevents the panel from closing by clicking outside the panel.

Summary

In this chapter, we learned more about how jQuery Mobile enhances basic HTML to provide additional layout controls to our mobile pages. With dialogs, we learned how to provide a basic, quick, modal message to users. With grids, we learned a new way to easily layout content in columns. We learned about collapsible content blocks—a cool way to share additional content without taking up as much screen space. We looked at popups, responsive tables, and finally, panel controls.

In the next chapter, we demonstrate a full, real example that creates a basic Note Tracker. It makes use of additional HTML5 features, as well as some of the UI tips you've learned over the past few chapters.

8
Moving Further with the Notekeeper Mobile Application

In this chapter we're going to begin assembling everything we've learned about lists, forms, pages, and content formatting thus far into a usable "mobile application" — the Notekeeper application.

In this chapter, we will do the following:

- Accept user input using forms
- Store user-inputted data locally using the HTML5 localStorage functionality
- Demonstrate how to add, edit, and remove items from the page dynamically

What is a mobile application?

Before writing our first mobile application, perhaps we should define what one is. Wikipedia defines it as follows (http://en.wikipedia.org/wiki/Mobile_app):

> *A mobile application (or mobile app) is a software application designed to run on smartphones, tablet computers and other mobile devices.*

While it's true that jQuery Mobile apps are written in HTML, CSS, and JavaScript, that doesn't prevent them from being sophisticated pieces of software. They can certainly be developed with mobile devices in mind.

Some critics might note that it can't really be software unless it's "installed". As you'll see later in the book, jQuery Mobile applications can actually be installed on a wide array of devices (including iOS, Android, and Windows Mobile) when coupled with the open source library PhoneGap (`http://phonegap.com/`). This means that you'll be able to have your cake and eat it too. You might be asking yourself if code written using jQuery Mobile can be considered as software, and as you'll find out in this chapter, the answer is yes.

Designing your first mobile application

The goal of any piece of software is to meet a need. Gmail met a need by freeing users from a single computer and letting them check their e-mail from any web browser. Photoshop met a need by allowing users to manipulate photos in ways no one had ever done. Our Notekeeper application meets a need by allowing us to record simple notes for later reference. I know that this is sort of a letdown by comparison, but we've got to start somewhere, right?

When building software, it's a good idea to spend time up front writing out a specification for your project: what it will do, what it will look like, and what it should have. Remember that if you don't know what you're building, how will you ever know if it's done?

Listing out the requirements

We already know what we want our application to do—take notes. The problem is that there are so many ways that you could build a note-taking app that it's essential to sketch out just what we want ours to do. Not too much, not too little, but just enough...for now. It's a point of fact with developers that our applications are never "done", they're only finished "for now". With Notekeeper, we've decided that we want to be able to do the following things with our application:

- Add a note
- Display a list of notes
- View a note/delete a note

After deciding what tasks our app needs to accomplish, we need to decide how it will accomplish them. The easiest approach is to simply write those things out in a list. By breaking each part down into smaller pieces, we make it easier to understand, and to see just what we need to make it work. It's just like getting directions to your favorite restaurant; a left turn here, a roundabout there, and you're sitting down at the table before you know it. Let's look at each thing we want Notekeeper to do, with the help of the following pieces and parts:

- Adding new notes (form)
  - **A form container**: All user input widgets should be wrapped and contained within a form
  - **A title**: This would be the name of the note and will also be used to display the existing notes
  - **The note itself**: The content or body of the note
  - **A Submit button**: This will trigger the saving of the note

- Displaying a list of existing notes (listview)
  - **A row item containing the title of the note**: This row should be a link to a page containing the body of the note
  - **A section header row**: This might be nice feature

- Viewing note details (label, paragraph, button)
  - A label for the title of the note
  - A paragraph containing the content of the note
  - A button labeled `Delete`
  - A Back button

Building your wireframes

Now that we've listed out the functionalities for our app, how about we sketch each piece so that we get an idea of what we want it to look like? Don't worry if you failed art or if you can't draw a stick figure. Use a ruler if you have to, or consider using Microsoft Excel, or PowerPoint if you have those. You just need to be able to draw some boxes and text labels. There are a number of free or inexpensive tools you can use for this purpose. A popular wireframing tool that runs in the browser is **Balsamiq Mockups** (`http://balsamiq.com/`).

Designing the add note wireframe

Now, what about the add note part? We decided that it needs a title, a box for the note, and a Submit button. The form is an invisible container so we don't need to draw that:

Display notes wireframe

The listview is an integral part of mobile development. It's the simplest way to group similar items together, plus, it offers lots of extra functionality, such as scrolling and built-in images for links. We'll be using a listview to display our list of notes:

View note/delete button wireframe

Finally, once we've added a note, we need to be able delete the evidence; I mean, clear out old notes to make way for new ones. Note that we've also sketched out a back button. Once you start seeing things laid out, you'll find that you've forgotten something really important (such as being able to return to the previous page):

Writing the HTML

Now that our wireframes are done and we're happy with them, it's time to turn pencil drawings into 1s and 0s. Because our app is relatively simple, none of the HTML should be difficult. You're more than halfway through the book after all and you should be able to do these things in your sleep.

The HTML that you come up with should look remarkably like what's shown in the following code snippet. Let's examine it together:

```
Code 8-1: notekeeper.html
<!DOCTYPE html>
<html>
  <head>
    <title>Notekeeper</title>
    <meta name="viewport" content="width=device-width, initial-
      scale=1">
    <link rel="stylesheet" href="http://code.jquery.com/mobile/
      latest/jquery.mobile.min.css" />
    <script src="http://code.jquery.com/jquery-1.8.2.js"></script>
    <script src="http://code.jquery.com/mobile/latest/
      jquery.mobile.min.js"></script>
    <script src="application.js"></script>
  </head>
  <body>
    <div data-role="page">
    <div data-role="header">
      <h1>Notekeeper</h1>
    </div>
      <div data-role="content">
        <form>
          <div>
            <input id="title" type="text" placeholder="Add a note" />
          </div>
```

```
      <div>
        <textarea id="note" placeholder="The content of your
          note"></textarea>
      </div>
      <div class="ui-grid-a">
      <div class="ui-block-a">
      <input id="btnNoThanks" type="submit" value="No Thanks" />
    </div>
      <div class="ui-block-b">
        <input id="btnAddNote" type="button" value="Add Note" />
      </div>
    </div>
  </form>
  <ul id="notesList" data-role="listview" data-inset="true">
    <li data-role="list-divider">Your Notes</li>
    <li id="noNotes">You have no notes</li>
  </ul>
    </div>
    <div data-role="footer" class="footer-docs">
      <h5>Intro to jQuery Mobile</h5>
    </div>
    </div>
  </body>
</html>
```

Our Notekeeper application will make use of a single HTML file (notekeeper.html)
and a single JavaScript file (application.js). Up until this point, none of the code
you've written has really needed JavaScript, but once you begin writing more complex
applications, JavaScript will be a necessity. Preview the HTML from Code 8-1 in your
web browser and you should see something similar to the following screenshot:

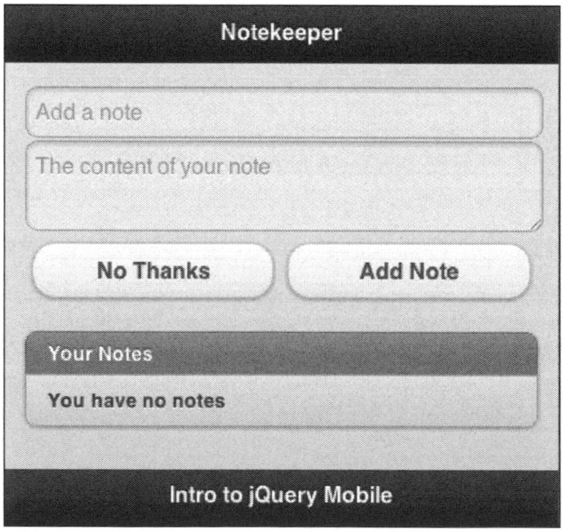

Notice that we're displaying the **Add Note** form on the same page as the form to view notes. With mobile application development, it's a good idea to condense things where possible. Don't make this a hard and fast rule, but because there's so little to our app, it's an acceptable decision to place both parts together as long as they're clearly labeled. You can see that this page meets all the requirements we set for adding a note and for displaying our existing notes. It has a title input field, a note input field, a save button, and the entire thing is wrapped inside a form container. It also has a listview that will be used to display our notes once we start adding them. What isn't seen here is a Delete button, but that will show up once we add our first note and view the details page.

Adding functionalities with JavaScript

As already mentioned in this book, you don't need to write any JavaScript to get your money's worth from jQuery Mobile. As you progress in your experience with jQuery Mobile, you'll begin to see how much additional value JavaScript can add to your projects. Before we look at the code, let's talk about how it will be structured. If you've done any web designing or development earlier, you've probably seen JavaScript. It has been around since 1995. The problem is that there's been many different ways to do the same thing in JavaScript and not all of them are good.

The JavaScript code in this application will use what's called a **design pattern**. It's just a fancy term that specifies a certain structure to the code. There are the following three main reasons for using an existing design pattern:

- It helps our code stay organized and tidy.

- It prevents the variables and functions we write from being accidentally overwritten or altered by any other code we might add, for example, a jQuery plugin perhaps, or code that's being loaded in from a third-party website.

- It will help future developers acclimatize themselves to your code much more rapidly. You are thinking about future developers as you work on the next Facebook killer, right?

Let's take a look at a very simple implementation of this design pattern before jumping into the complete code:

```
Code 8-2: kittyDressUp.js
$(document).ready(function(){
    // define the application name
    var kittyDressUp = {};
    (function(app){
        // set a few variables which can be used within the app
        var appName = 'Kitty Dress Up',
```

```
                version = '1.0';
          app.init = function(){
              // init is the typical name that developers give for the
              // code that runs when an application first loads
              // use whichever word you prefer
              var colors = app.colors();
          };
          app.colors = function(){
              var colors = ['red','blue','yellow','purple'];
              return colors;
          };
          app.init();
    })(kittyDressUp);
});
```

If you're familiar with JavaScript or jQuery, you'll probably see some elements that you recognize. For those readers who aren't familiar with jQuery or JavaScript, we'll review this example line by line. `KittyDressUp.js` starts off with jQuery's most recognizable element: `$(document).ready()`. Any code contained within this block waits to execute until the document or the HTML page is completely loaded. This means that you, the developer, can be assured that everything that needs to be on the page is there before your code runs:

```
$(document).ready({
  // I'm ready captain!
});
```

In simple terms, the next line creates a variable named `kittyDressUp` and assigns it a value of an empty object. However, in our code, this new object will contain our entire application:

```
// define the application name
var kittyDressUp = {};
```

The following declaration is the core of the *Kitty Dress Up* application. It creates a function that accepts a single argument and then immediately calls itself, passing in the empty object we created in the previous line of code. This is known as a **self-executing function** and is what keeps the external code from interfering with our application:

```
(function(app){
  // define the app functionality
})(kittyDressUp);
```

The next two lines set a few variables that can only be accessed from within the context or scope of our application:

```
// set a few variables which can be used within the app
var appName = 'Kitty Dress Up',
    version = '1.0';
```

Finally, the last few lines set up two functions that are available for use within the application. You can see that each function is assigned a name that is within the scope of the larger application. This is known as a **namespace**. The app variable is where the function lives, and the word after the dot (.) is the function name. Notice that, within the init function, we're calling another function contained within the same application called app.colors(). We could also reference any of the variables that we defined at the top as well:

```
app.init = function(){
    // init is the typical name that developers give for the
    // code that runs when an application first loads
    // use whatever word you prefer
    var colors = app.colors();
}
app.colors = function(){
    var colors = ['red','blue','yellow','purple'];
    return colors;
}
app.init();
```

Remember that app was the name of the parameter passed into the self-executing function and that its value is an empty object. Taken as a whole, these few lines create an object named kittyDressUp that contains two variables (appName and version), and two functions (init and colors). This example, as well as the code for Notekeeper, are simple examples, but they illustrate how to go about wrapping up code for various pieces of a larger app into discrete packages. In fact, after kittyDressUp.js runs, you could even pass the kittyDressUp object into yet another set of code for use there.

Phew…everyone take five, you've earned it.

Storing Notekeeper data

Now that we're back from our five-minute break, it's time to roll up our sleeves and get to work adding functionalities to our app. While we've talked about how we want Notekeeper to behave, we haven't discussed the core issue of where to store the note data. There are a few possibilities, all of which have pros and cons. Let's list them out:

- **Database (MySQL, SQL Server, and PostgreSQL)**: While a database would be the ideal solution, it's a little complex for our app because it requires Internet connectivity, and you'd need a server-side component (such as ColdFusion, PHP, and .NET) acting as a middleman to save notes to the database.

- **Text file**: Text files are great because they take up very little room. The problem is that as a web app, Notekeeper can't save files to the user's device.

- **localStorage**: `localStorage` is relatively new, but it's quickly becoming a good option. It stores information on the user's machine in key/value pairs. Although it's got a size limit, it's pretty large for plain text, most modern browsers support it, and it can also be used in the offline mode.

Using localStorage

For the purposes of this chapter, we'll be selecting `localStorage` as our method of choice. Let's take a quick look at how it behaves so that you'll be familiar with it when you see it. As mentioned previously, `localStorage` works on the premise of storing data in key/value pairs. Saving a value to `localStorage` works in one of two ways and is easy, no matter which one you choose:

```
localStorage.setItem('keyname','this is the value I am saving');
```

Or

```
localStorage['keyname'] = 'this is the value I am saving';
```

Which version you choose is personal preference, but we'll be using the second method, square brackets, because it requires slightly less typing. One issue we'll run into is that `localStorage` can't directly store complex data such as arrays or objects. It only stores strings. That's a problem because we're going to store all of our data inside one variable so that we always know where it lives. Never fear, we can pull a fast one on `localStorage` and convert our complex object into a string representation of itself using a built-in function called `stringify()`. This conversion process, from complex object to string, is called **serialization**.

The following code snippet shows how it works:

```
// create our notes object
var notes = {
  'note number one': 'this is the contents of note number one', 'make
    conference call': 'call Evan today'
  }
// convert it to a string, then store it.
localStorage['Notekeeper'] = JSON.stringify(Notekeeper);
```

Retrieving a value is just as simple as setting it, and it also offers two options. You'll usually want to define a variable to receive the content of the localStorage variable:

```
var family = localStorage.getItem('my family');
```

Or

```
var family = localStorage['my family'];
```

If you're retrieving a complex value, there's an additional step that must be performed before you can use the content of the variable. As we just mentioned, to store complex values you must first use the stringify() function, which has a counterpart function called parse(). The parse() function takes the string containing that complex object and turns it back into pure JavaScript, a process called **deserialization**. It's used as follows:

```
var myFamily = ['andy', 'jaime', 'noelle', 'evan', 'mason', 'henry'];
localStorage['family'] = JSON.stringify(myFamily);
var getFamily = JSON.parse(localStorage['family']);
```

Finally, if ever you want to delete the key completely, you can accomplish it in a single line of code, again with two flavors:

```
localStorage.removeItem('family');
```

Or

```
delete localStorage['family'];
```

It's worth noting that if you try to retrieve a key that doesn't exist within localStorage, JavaScript won't throw an error. It'll just return "undefined," which is JavaScript's way of saying "sorry, but nothing's there". The following code snippet is an example:

```
var missing = localStorage['yertl the turtle'];
console.log(missing);
// returns undefined
```

Effective use of boilerplates

There is one last thing before we start building our JavaScript file. In our application, we're only going to have one JavaScript file, and it's going to contain the entire codebase. This is fine for smaller apps like ours, but it's a bad idea for larger apps. It's better to break up your project into distinct pieces, then put each of those into their own files. This makes it easier for teams of developers to work together (for example, Evan works on the login process, while Henry builds out the list of vendors). It also makes each file smaller and easier to understand because it only addresses one part of the whole. When you want all of the pieces of your app to have a similar structure and design, it's a good idea to start each section with a **boilerplate**. We'll be using a boilerplate for our app's only file (which you can see in the following code snippet, Code 8-3). You might notice it looks very similar to the kittyDressUp example, and you'd be right:

```
Code 8-3: application.js
$(function(){
  // define the application
  var Notekeeper = {};
  (function(app){
    // variable definitions go here
    app.init = function(){
      // stuff in here runs first
    }
    app.init();
  })(Notekeeper);
});
```

Building the Add Note feature

At last, we can get started building! As it's difficult to display a list of notes that don't exist, much less delete one, we'll start writing the Add Note functionality first. For a user to be able to add a note, they have to enter a title, the content of a note, then hit the Submit button. So let's start there.

Adding bindings

We're going to create a new, empty, function block under the app.init() function definition. It should look something similar to the following line of code:

```
app.bindings = function(){
}
```

The `bindings` function can contain any piece of code that needs to fire when a user does something in our app, such as clicking on the Submit or the Delete button. We group that code together for the sake of organization. Within the `bindings()` function, we're going to add the following lines. This will fire when a user clicks on the Submit button on the Add Note form:

```
// set up binding for form
$('#btnAddNote').on('click', function(e){
  e.preventDefault();
  // save the note
  app.addNote(
    $('#title').val(),
    $('#note').val()
  );
});
```

jQuery's `val()` function is a shorthand method used to get the current value of any form input field.

The following are a few notes about this new addition:

*   When using jQuery, there will always be more than one way to accomplish something, and in most cases, you simply pick the one that you like the best (they usually offer identical performance). You might be more familiar with `$('#btnAddNote').click()`, and that's just fine as well.

*   Notice that the `.on()` method accepts a single parameter called `e` which is the event object (in this case, a click event). We call `e.preventDefault()` to stop the standard click event from happening on this element, but still allow the remaining code to continue running. You might have seen other developers use `return false`, but jQuery best practices recommend using `e.preventDefault()` instead.

*   Within the click binding, we're calling the `addNote` function and passing the title typed in by the user and the note into it. The whitespace is unimportant, serving merely to make it easier to see what we're doing.

Even though we've added the binding to our code, if you run the app right now, nothing will happen when you click on the **Add Note** button. The reason is that nothing has actually called the `bindings()` function yet. Add the following line inside the `init()` function and you'll be ready to go:

```
app.init = function(){
  app.bindings();
}
```

Collecting and storing the data

Next, we add another new, empty, function block under `app.bindings`:

```
app.addNote = function(title, note){
}
```

Now, because we're storing all of our notes into one key within `localStorage`, we first need to check if any notes already exist. Retrieve the Notekeeper key from `localStorage`, save it to a variable, then compare it. If the value of the key we ask for is an empty string or is undefined, we'll need to create an empty object. If there is a value, we take that and use the `parse()` function to turn it into JavaScript:

```
var notes = localStorage['Notekeeper'];
if (notes == undefined || notes == '') {
  var notesObj = {};
} else {
  var notesObj = JSON.parse(notes)
}
```

Notice that we're expecting two variables to be passed into the `addNote()` function, `title` and `note`. Next, we replace any spaces in the title with dashes; this makes it easier for some browsers to understand the string of text. Then we place the key/value pair into our newly minted `notes` object:

```
notesObj[title.replace(/ /g,'-')] = note;
```

The JavaScript `replace` method makes string manipulation quite simple. It acts on a string, taking a search term and a replacement term. The search term can be a simple string or a complex regular expression.

The next step is to take our `notesObj` variable, pass it into `stringify()` and place the return value into `localStorage`. We then clear the values from the two input fields to make it easier for the user to input another note. As a rule in building software, it's a nice touch to return the interface to its original state after an action, such as adding or removing content:

```
localStorage['Notekeeper'] = JSON.stringify(notesObj);
// clear the two form fields
$note.val('');
$title.val('');
//update the listview
app.displayNotes();
```

All of these variable definitions should be familiar to you with perhaps one exception that we should point out. Many jQuery developers like to use conventional naming for variables that contain jQuery objects. Specifically, they prepend the variable name with a $ sign just like with jQuery. This lets them, or future developers know what's contained within the variable. Let's go ahead and add those definitions to the top of our app. Just after the line that reads // variable definitions go here, add the following lines. They refer to the title input field and the note text area field respectively:

```
var $title = $('#title'),
    $note = $('#note');
```

As a final step to this function, we fire off a call to app.displayNotes() to update the list of notes. Since that function doesn't exist yet, let's create it next.

Building the Display Notes feature

You probably tested out the Add Note feature while in the previous section. This means that you'll have at least one note saved in localStorage for use in testing the Display Notes feature. By now, you'll be familiar with our first steps for any new section. Go ahead and add your empty displayNotes() function to hold our code:

```
app.displayNotes = function(){
}
```

Next, we need to retrieve all of our notes from localStorage:

```
// get notes
var notes = localStorage['Notekeeper'];
// convert notes from string to object
return JSON.parse(notes);
```

You might start to see a pattern with many of our functions; almost all of them begin with us retrieving notes from localStorage. While there are only two lines of code needed to perform this task, there's no need for us to repeat those two lines each time we need to get the notes. So we're going to write a quick helper function containing those two lines. It looks similar to the following code snippet:

```
app.getNotes = function(){
  // get notes
  var notes = localStorage['Notekeeper'];
  // convert notes from string to object
  return JSON.parse(notes);
}
```

With our new helper function in place, we can use it in the `displayNotes()` function as shown in the following code snippet:

```
app.displayNotes = function(){
  // get notes
  var notesObj = app.getNotes();
}
```

Now that we have the `notesObj` variable containing our packet of notes, we need to loop over that packet and output the contents:

```
// create an empty string to contain html
var notesObj = app.getNotes(),
      html = '',
      n; // make sure your iterators are properly scoped
// loop over notes
for (n in notesObj) {
  html += li.replace(/ID/g,n.replace(/-/g,' ')).replace(/LINK/g,n);
}
$ul.html(notesHdr + html).listview('refresh');
```

It might seem odd for the line inside the `for` loop to have multiple `replace` statements, but the nature of JavaScript allows for methods to be chained. Chaining allows us to run multiple operations (on the same element) within a single statement. Adding an additional method call simply repeats the process.

There might be some new concepts in this code block, so let's take a closer look. The variable named `html` is nothing special, but the way we're using it might be. As we loop over the existing notes, we're storing new information into the `html` variable along with whatever else is inside it. We accomplish this by using the `+=` operator that allows us to assign and append at the same time.

The second thing you might notice is the `li` variable on the right-hand side of the assignment. Where does that come from? That's a template for a single list item that has not yet been created. Let's do that now before we talk about it. At the top of your `app.js` file, just after the line that reads `// variable definitions go here`, add the following lines of code:

```
var $title = $('#title'),
    $note = $('#note'),
    $ul = $('#notesList'),
    li = '<li><a href="#pgNotesDetail?title=LINK">ID</a></li>',
```

You'll already be familiar with the convention of adding a $ symbol before a variable to indicate a jQuery object. That's what we're doing with the $ul variable. The second new variable, the li, is slightly different. This contains the HTML for a single list item that will display a note's title. It's best practice to avoid mixing HTML or CSS in with your JavaScript wherever possible. We're declaring this as a template now, in case we decide to use it in multiple places later.

The other part that might be of interest is the way we're using the li variable. When calling the string replace function, we're looking for all occurrences of the word LINK (uppercase intended) and replacing it with the title of the note. Because JavaScript is a case-sensitive language, it's a safe assumption that we won't run into a natural occurrence of that word.

Dynamically adding notes to our listview

There's one final thing we need to put in place before our notes show up on the page. You might have noticed that the only place that calls the displayNotes() function appears within the addNote() function. This is a good place for it, but it can't be the only place. We need something that runs when the page first loads. The prime place for this would be in the init() function, and that's where we'll place it.

There's one problem though, we can't just load our notes and run; what happens if there are no notes? We need a nice message to display to the user so that they don't think something's wrong. Let's create a new function called app.checkForStorage() that handles all of this:

```
app.checkForStorage = function(){
    var notes = app.getNotes();
    // are there existing notes?
    if (!$.isEmptyObject(notes)) {
        // yes there are. pass them off to be displayed
        app.displayNotes();
    } else {
        // nope, just show the placeholder
        $ul.html(notesHdr + noNotes).listview('refresh');
    }
};
```

By now, all of this should be familiar to you: calling our `app.getNotes()` method for notes and calling the `displayNotes()` function if it finds them. The second part has some new items though. When we set the HTML for the `$ul` jQuery object, we're calling two new variables. One for the listview header, and another if we don't have any notes. Let's add those two variable definitions now. Under `// variable definitions go here`, add the following lines:

```
var $title = $('#title'),
    $note = $('#note'),
    $ul = $('#notesList'),
    li = '<li><a href="#pgNotesDetail?title=LINK">ID</a></li>'
    notesHdr = '<li data-role="list-divider">Your Notes</li>',
    noNotes = '<li id="noNotes">You have no notes</li>';
```

The last part of the line normally goes unnoticed, but we won't let it. It's really crucial. jQuery Mobile offers several options to developers. Apart from the option of having a static HTML code that's already on the page when it loads, jQuery Mobile also provides an option for adding HTML code on the fly. That really gives developers lots of flexibility, but it presents a unique challenge as well. By design, jQuery Mobile converts HTML into stylish looking widgets before the page is displayed to the user. This means that any HTML added after that will be presented to the user without any style.

However, jQuery Mobile also offers a way to get around this by building in the ability to refresh each and every element that it converts. Most of them have a built-in function corresponding to the name of the element; in our case, it's the `listview()` function. Actually, this method offers the ability to add a completely new listview to the page. In our situation, we only care about refreshing the one we have, so we simply add the `refresh` keyword and jQuery Mobile converts your plaintext listview. Try leaving that last part out and see just how much work jQuery Mobile saves you. Maybe you should add the jQuery Mobile team to your Christmas card list?

Finally, we have to actually call our newest function. Within the `init()` function, add the following line. Then reload the page and watch your notes load up:

```
app.checkForStorage();
```

Viewing a note

At this point, we should be able to create a new note and have that note be immediately displayed in our listview. In fact, the rows in the listview are already links, but they just don't work; let's change that right now.

Using the .on() method

Add the following lines to the `bindings()` function:

```
$(document).on('click', '#notesList a', function(e){
    e.preventDefault();
    var href = $(this)[0].href.match(/\?.*$/)[0];
    var title = href.replace(/^\?title=/,'');
    app.loadNote(title);
});
```

This new binding has a few new concepts, so let's unpack them. First up, we're not using the `bind()` function, instead, we use jQuery's `on()` function. The difference is that `bind()` only works on existing page elements, whereas `on()` is proactive. It works on existing elements as well as on the ones that get created after the binding is applied.

The second and third lines of the binding might look a little confusing, but they only do one thing. They retrieve the URL from the `href` attribute of the link that was clicked. The `li` template we defined earlier in the chapter contained the following URL for each list item:

```
#pgNotesDetail?title=LINK
```

After the `displayNote()` function runs, the URL looks like the following (run your mouse over each list item to see the link displayed at the bottom of your browser window):

```
#pgNotesDetail?title=the-title-of-the-note
```

Finally, we tell our code to run a new function appropriately named `app.loadNote()`.

Dynamically creating a new page

If you haven't already created the new empty function block for our new `loadNote()` function, go ahead and do it now. Remember that we're passing in the title of the note we want to view, so make sure to add that as an argument in the `loadNote()` function:

```
app.loadNote = function(title){
}
```

Then include the following lines at the top of the function:

```
// get notes
var notes = app.getNotes(),
    // lookup specific note
    note = notes[title],
```

```
        // define the "new page" template
        page = ['<div data-role="page" data-url="details" data-add-back-
    btn="true">',
                '<div data-role="header">',
                    '<h1>Notekeeper</h1>',
                    '<a id="btnDelete" href="" data-href="ID" data-
    role="button" class="ui-btn-right">Delete</a>',
                '</div>',
                '<div data-role="content"><h3>TITLE</h3><p>NOTE</p></
    div>',
            '</div>'].join(''),
```

The first line retrieves our `note` object, while the second line pulls the specific note that the user has requested. The third variable definition breaks the rule we mentioned earlier in the chapter about mixing HTML and JavaScript; but every rule has exceptions. We're defining it here as opposed to the header of our JS file, because this is the only place it is needed. This still serves the purpose of keeping the document organized.

The `page` variable now contains all of the HTML needed to display a "note details" page to the user. Do you recall that our app has only one HTML file? We're actually creating an entire page from scratch using the previous HTML code. There are also some details in it worth pointing out:

- By default, jQuery Mobile does not offer a Back button for pages. You can, however, enable one on a page-by-page basis using the `data-add-back-btn="true"` attribute on any `div` tag that also has a `data-role="page"` attribute.

- The `data-url` attribute is an identifier used by jQuery Mobile so that it can keep track of multiple pages that are generated.

- This approach to string concatenation might look different to you. When you only have a few words, it's okay to say `'andy' + 'jaime'` but when you have multiple lines, a nice trick is to create an array where each line is an item in the array. Then you join each item together and presto! It has the added benefit of keeping things neat and tidy too.

Now that we have a whole page contained within a variable, what can we do with it? The first thing we can do is to turn it into a jQuery object. By wrapping any distinct chunk of HTML with `$()`, you turn it into a Grade-A jQuery object:

```
newPage = $(page);
```

Then we can take the HTML of that newly created page and replace parts of it with the values from our selected note:

```
// append it to the page container
newPage.html(function(index,old){
  return old
    .replace(/ID/g,title)
    .replace(/TITLE/g,title
    .replace(/-/g,' '))
    .replace(/NOTE/g,note)
}).appendTo($.mobile.pageContainer);
```

Since Version 1.4, jQuery has offered the option of a callback within certain functions. These include `.html()`, `.text()`, `.css()`, and a few others. This function expects two arguments, of which the second contains the full HTML currently contained within the matching element. This means that we can make tweaks to the HTML contained inside our `newPage` variable without having to completely change it. Wonderful, isn't it?

Next, we're appending the entire `newPage` variable to the end of the current page, referenced here by the `$.mobile.pageContainer` constant. Finally, because we cancelled the default click action in our binding, we have to tell the link to perform an action that is to forward the user to this newly created page. jQuery Mobile offers a built-in way to do this:

```
$.mobile.changePage(newPage);
```

And now for the grand reveal. If you load up `notekeeper.html` in your browser, you should be able to add, display, and finally view notes, all within the confines of a single browser window. Isn't jQuery Mobile great?

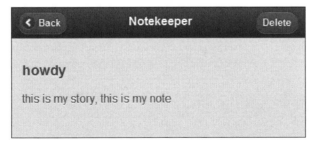

Deleting a note

Looking back to the requirements for our app, we're doing pretty well. We've written HTML code that sets up the document structure, allows us to add a note, display notes, and view a note. All that's left is deleting a note and it begins with a last binding setup in our `bindings()` function. Let's add it right now:

```
$(document).on('click', '#btnDelete', function(e){
    e.preventDefault();
    var key = $(this).data('href');
    app.deleteNote(key);
});
```

There's only one item that might be new to you in this binding—the use of jQuery's `.data()` function. HTML5 allows you to store arbitrary data directly on any HTML element by using an attribute prepended with `data-` and this ability is at the core of jQuery Mobile's functionality. Anywhere you see `data-role="something"`, you're seeing HTML5 data in action. Furthermore, jQuery allows you to retrieve any `data-` value by using the `.data()` function and passing in the key of the item you want to view. In the preceding case, we've stored the title of the note into a `data-href` attribute directly on the Delete button within the view page. Because the binding we're adding is a click handler assigned to the Delete button, we can retrieve the title of the note by calling `$(this).data('href')`. Neat-o!

This will be the last function that we add in this chapter. Are you sad? It's a poignant moment for certain, but we can look back on this with fondness after you're a successful jQuery Mobile developer. Once again, we start with an empty function that accepts a single argument, the title of the note we're deleting:

```
app.deleteNote = function(key){
}
```

Follow the function definition up with our helper function for retrieving notes:

```
// get the notes from localStorage
var notesObj = app.getNotes();
```

Then we delete the note. You've already seen this in action when we reviewed `localStorage`, so it should be familiar to you:

```
// delete selected note
delete notesObj[key];
// write it back to localStorage
localStorage['Notekeeper'] = JSON.stringify(notesObj);
```

Deleting the note is followed in quick succession by writing the remaining notes back to `localStorage`. The final two lines in the `deleteNote()` function take us back to the main page of the app—the list of notes. They also trigger the original `checkForStorage()` function:

```
// return to the list of notes
$.mobile.changePage('notekeeper.html');
// restart the storage check
app.checkForStorage();
```

The last line may seem odd to you, but keep in mind that we don't know in advance if there are still any notes left. Running through the storage check allows us to display the placeholder text, in case there are no notes. It's a good habit to get into, as it helps our app become less prone to errors.

Summary

In this chapter, we built a living, breathing mobile application with jQuery Mobile. Stop and give yourself a pat on the back. We walked through the process of listing the requirements for our app, building the wireframes, and writing the HTML. We learned about HTML5's `localStorage`, using templates for text replacement, and some of the cooler features of jQuery Mobile, including dynamically adding and refreshing elements on the page.

In the next chapter, you'll learn how to set global configuration options for jQuery Mobile, how to use other utility APIs within jQuery Mobile to streamline your code, and take advantage of the great work the jQuery Mobile team has already done on your behalf.

9

jQuery Mobile Configuration, Utilities, and JavaScript Methods

In this chapter, we will look at how JavaScript can be used to further configure and enhance jQuery Mobile websites. So far we've made use of HTML and CSS to generate everything. Now we'll look at additional scripting that adds additional functionalities to your sites.

In this chapter, we will do the following:

- Explain how jQuery Mobile sites can be configured via JavaScript
- Discuss the various JavaScript utilities that ship with jQuery Mobile and how they can be used
- Explain the APIs used to work with the enhanced jQuery Mobile form and widget controls

Configuring jQuery Mobile

jQuery Mobile does many things for you—from improving page navigation to changing how form controls work. All of this is done in an effort to make your content work better in a mobile environment. There will be times, however, when you do not want jQuery Mobile to do something, or you perhaps simply want to slightly tweak how the framework acts. That's where configuration comes in.

To configure a jQuery Mobile site, you begin by writing code that listens for the `mobileinit` event. This can be listened to using a normal jQuery event handler, similar to the following code snippet:

```
$(document).bind("mobileinit", function() {
  //your customization here
});
```

In order for this event to be captured, you must configure it before jQuery Mobile is actually loaded. The simplest way to do this, and the way recommended by the jQuery Mobile docs, is to simply place this code in a script that is loaded before the jQuery Mobile JavaScript library. The following code snippet shows what the header of our files typically looks like:

```
<!DOCTYPE html>
<html>
  <head>
    <title>Dialog Test</title>
    <meta name="viewport" content="width=device-width, initial-
      scale=1">
    <link rel="stylesheet" href="http://code.jquery.com/mobile/
      1.3.2/jquery.mobile-1.3.2.min.css" />
    <script src="http://code.jquery.com/jquery-
      1.9.1.min.js"></script>
    <script src="http://code.jquery.com/mobile/
      1.3.2/jquery.mobile-1.3.2.min.js"></script>
  </head>
```

Notice that the jQuery Mobile library is the last one loaded. We can simply add in a new script tag before it:

```
<!DOCTYPE html>
<html>
  <head>
    <title>Dialog Test</title>
    <meta name="viewport" content="width=device-width, initial-
      scale=1">
    <link rel="stylesheet" href="http://code.jquery.com/mobile/
      latest/jquery.mobile.min.css" />
    <script src="http://code.jquery.com/jquery-
      1.9.1.min.js"></script>
   <script src="config.js"></script>
    <script src="http://code.jquery.com/mobile/
      latest/jquery.mobile.min.js"></script>
  </head>
```

Configuring jQuery Mobile is as simple as updating the $.mobile object. The following code snippet is a simple example:

```
$(document).bind("mobileinit", function() {
  $.mobile.someSetting="some value here";
});
```

This object contains a set of key/value pairs for the various settings that can be configured. You don't actually create it; it exists when the event handler is run. Another option is to make use of jQuery's extend() functionality, as shown in the following code snippet:

```
$(document).bind("mobileinit", function() {
  $.extend($.mobile, {
    someSetting:"some value here"
  });
});
```

Either form is ok and works exactly the same. Use whichever you feel is more comfortable. Now, let's look at the various configuration options:

Settings	Use
ns	This is the namespace value used for data attributes. It defaults to nothing. You would specify a value here if you wanted to prefix the jQuery Mobile-recognized data attributes. So for example, if you wanted to use data-jqm-role="page" instead of data-role="page", you would configure the ns value to be jqm.
activeBtnClass	This simply sets the class name used for buttons in the active state. The default for this value is ui-btn-active.
activePageClass	This sets the class name for pages that are currently being viewed. The default for this value is ui-page-active.
ajaxEnabled	We've discussed earlier how Ajax is used for both page loads and form submissions. If you wish to disable this, set this value to false. The default, obviously, is true.
allowCrossDomainPages	This is a security setting that defaults to false; setting this to true allows for remote pages to be loaded via $.mobile.loadPage. This is normally only required for PhoneGap applications that load content from another server.

Settings	Use
autoInitializePage	Normally, jQuery Mobile will run $.mobile.initializePage on load. This displays the renders page. (At this time, this particular function isn't properly documented.) If you wish to disable this default value, set the value of autoInitializePage to false. You will need to run $.mobile.initializePage manually.
buttonMarkup.hoverDelay	Specifies a delay value to buttons for triggering the hover and down classes. The lower the value, the quicker the button will appear "depressed" when touched on the mobile device. It defaults to 200.
defaultDialogTransition	Specifies what transition should be used to show or hide dialogs. The default is pop. Possible values are fade, flip, pop, slide, slidedown, and slideup.
defaultPageTransition	Like the previous option, this setting is used for transitions; this time, for page loads. The default is fade and the options similar to the previous option are possible.
dynamicBaseEnabled	Used to signal if a dynamic base tag is used. This defaults to true and should only be set to false if you're using another web framework that requires a particular base tag reference.
gradeA	Used to determine what actually constitutes a "good" browser. This is handled by jQuery Mobile, but if you want to overrule the framework or define some other condition that must be met, you would need to provide a function here that returns a Boolean (true or false) value.
getMaxScrollForTransition	One of the performance tricks jQuery Mobile uses is to automatically disable page transitions if you're navigating from or to a page that is *very* long. So for example, imagine you are at the bottom of a very long page and you click on a link to load a new page. If jQuery Mobile determines that you've scrolled to a value of three times the window height, it will disable the transition. That value, 3, can be configured by this property.
hashListeningEnabled	Refers to the ability to listen to changes in the location.hash property of the browser. jQuery Mobile handles this normally, but if the value is set to false, you can write your own code to respond to these changes.

Settings	Use
`ignoreContentEnabled`	Normally, jQuery Mobile automatically enhances everything it can. You can disable this in some cases at a control level, but you can also tell jQuery Mobile to ignore everything within a particular container by adding `data-enhance=true`. If you make use of this feature, your configuration must set `ignoreContentEnabled` to `true`. This tells jQuery Mobile to look for, and respect, that particular flag. This is set to `false` by default, and allows jQuery Mobile to work its magic quite a bit faster.
`linkBindingEnabled`	jQuery Mobile typically listens to all link clicks. If you wish to disable this globally, you can do so with this setting.
`maxTransitionWidth`	This is used to set a max size for transitions; if set to a value that is smaller than a window that it will be transitioned *into*, no visual transitions will be used.
`minScrollBack`	jQuery Mobile will attempt to remember your scrolled position in a page when you return to it. This can be useful on a large page that the user returns to after visiting another page. By default, the scroll will be remembered if it is more than `150`, the default.
`pageLoadErrorMssage`	This is a message shown to users if an error occurs when loading a page. The default is **Error Loading Page**, but could be changed for localization reasons. (Or any reason, really).
`pageLoadErrorMessageTheme`	This is the theme to use when a page load error dialog is displayed. The default is `e`.
`phonegapNavgiationEnabled`	If enabled, PhoneGap's navigation helper will be used when sending the user to their previous location. This was added to help with issues under Android. The default is `false`.
`pushStateEnabled`	This tells jQuery Mobile to use the HTML5 `pushState` functionality instead of hash-based changes for page navigation. This defaults to `true`.

Settings	Use
subPageUrlKey	jQuery Mobile supports multiple pages within one file. In order to make these "virtual" pages bookmarkable, jQuery Mobile will append a value to the URL containing the prefix ui-page. For example, ui-page=yourpage. This setting lets you customize the prefix.
transitionFallbacks	This is not a simple setting, but rather a hashmap that allows you to specify a fallback for a transition that is not supported on a device. So for example, you can specify that the fallback for the slide transition is pop by using transitionFallsback["slide"] = "pop".

That's quite a few options, but typically, you will only need to configure one or two of these settings. Let's look at a simple example where a few of these are put to use. Code 9-1 is the home page for the application. Note the use of the additional script tags to load in our configuration:

```
Code 9-1: test1.html
<!DOCTYPE html>
<html>
  <head>
    <title>Page Transition Test</title>
    <meta name="viewport" content="width=device-width, initial-
      scale=1">
<link rel="stylesheet" href="http://code.jquery.com/mobile/1.3.2/
jquery.mobile-1.3.2.min.css" />
<script src="http://code.jquery.com/jquery-1.9.1.min.js"></script>
<script src="config.js"></script>
<script src="http://code.jquery.com/mobile/1.3.2/jquery.mobile-
1.3.2.min.js"></script>
  </head>
  <body>
    <div data-role="page" id="first">
      <div data-role="header">
        <h1>Dialog Test</h1>
      </div>
      <div data-role="content">
        <p>
          <a href="#page2">Another Page</a><br/>
          <a href="test2.html">Yet Another Page</a><br/>
        </p>
```

```
        </div>
      </div>
      <div data-role="page" id="page2">
        <div data-role="header">
          <h1>The Second</h1>
        </div>
        <div data-role="content">
          <p>
            This is the Second. Go <a href="#first">first</a>.
          </p>
        </div>
      </div>
    </body>
</html>
```

The file contains two jQuery Mobile pages and links to another page in test2.html. That page simply provides a link back so will not be included in the text. Now let's look at config.js:

```
Code 9-2: config.js
$(document).bind("mobileinit", function() {
  $.mobile.defaultPageTransition = "none";
});
```

In config.js, one setting is modified—the default page transition.

In an earlier chapter, we discussed forms and how jQuery Mobile automatically enhances controls. While you can suppress this enhancement on a control within your HTML, you can also tell jQuery Mobile a list of controls never to enhance. To set this list, specify a value for $.mobile.page.prototype.options.keepNative. The value should be a list of selectors. Any field that matches one of the selectors will *not* be enhanced. (As a reminder, you can disable the form field's auto-enhancement in HTML by adding data-role="none" to your form field.) Code 9-3 demonstrates an example of this:

```
Code 9-3: form.html
<!DOCTYPE html>
<html>
<head>
<title>Page Transition Test</title>
<meta name="viewport" content="width=device-width, initial-scale=1">
<link rel="stylesheet" href="http://code.jquery.com/mobile/1.3.2/
jquery.mobile-1.3.2.min.css" />
<script src="http://code.jquery.com/jquery-1.9.1.min.js"></script>
<script src="config2.js"></script>
<script src="http://code.jquery.com/mobile/1.3.2/jquery.mobile-
```

```
1.3.2.min.js"></script>
</head>

<body>

<div data-role="page" id="first">

    <div data-role="header">
       <h1>Form Test</h1>
    </div>

    <div data-role="content">

       <div data-role="fieldcontain">
           <label for="name">Name:</label>
           <input type="text" name="name" id="name" value=""  />
       </div>

       <div data-role="fieldcontain">
           <label for="email">Email:</label>
           <input type="text" name="email" id="email" value=""
class="boring"  />
       </div>

    </div>

</div>

</body>
</html>
```

Notice two things in particular. First, we're loading a config file in the header of the page. Second, our form has two text fields, but we've added a class to the second one named boring. Code 9-4 is our configuration file:

```
Code 9-4: config2.js
$(document).bind("mobileinit",  function() {
    $.mobile.page.prototype.options.keepNative = "input.boring";
});
```

We've specified that we want input tags with the class of `boring` to *not* be enhanced. The result is shown in the following screenshot:

Using jQuery Mobile utilities

Now that we've covered jQuery Mobile configuration, let's take a look at the utilities available to your applications. These are utilities provided by the framework and can be used in any application. You may not need all of them (or any) on your site, but knowing they are there can help save time in the future.

Page methods and utilities

Let's begin looking at methods and utilities related to pages and navigation between pages:

- `$.mobile.activePage`: This property is a reference to the current page.

- `$.mobile.changePage(page,options)`: This method is used to switch to another page. The first argument, `page`, can be either a string (the URL), or a jQuery DOM object. The `options` argument is an optional object of key/value pairs. These options are as follows:

    - `allowSamePageTransition`: Normally jQuery Mobile will not allow you to transition to the same page, but if set to `false`, this will be allowed.

    - `changeHash`: This determines if the URL should change.

    - `data`: This is either a string or an object of values passed to the next page.

    - `dataUrl`: This is the value used for the URL in the browser and is normally set by the page the user is being sent to. You can override this here.

    - `pageContainer`: jQuery Mobile will place pages within a DOM item that acts as a "bag" for all the pages. You can bypass this automatic collection and use another item in the DOM instead.

    - `reloadPage`: If a page already exists in the browser, jQuery Mobile will fetch it from memory. Setting this to `true` will force jQuery Mobile to reload the page.

    - `reverse`: This determines the "direction" of the transition.

    - `role`: jQuery Mobile will typically look for the `data-role` attribute of the page loaded. To specify another role, set this option.

    - `showLoadMsg`: Normally, jQuery Mobile shows a loading message when a page is fetched. You can disable this by setting this value to `false`.

    - `transition`: This specifies what transition to use. Remember that this can be configured at a global level as well.

    - `type`: We mentioned earlier that jQuery Mobile loads in new pages via an Ajax-based request. The `type` option allows you to specify the HTTP method used to load the page. The default is `get`.

- `$.mobile.loadPage(page,options)`: This is a lower-level function used when `$.mobile.changePage` is passed a string URL to load. Its first argument is the same as `$.mobile.changePage`, but its options do not include `dataUrl`. Those options are the same as those listed in the previous option, except for `loadMsgDelay`. This value gives time for the framework to try to fetch a page via the cache first.

- `$.mobile.navigate(url, data)`: Both `changePage` and `loadPage` are available to developers, but `navigate` may be a simpler way of doing the same thing — sending the user to a new location in your application. Simply provide a URL and an optional set of data. jQuery Mobile handles using the changes in the HTML5 History API if possible with the current browser.

In `Code 9-5`, a simple example of `$.mobile.changePage` is demonstrated:

```
Code 9-5: test3.html
<div data-role="page" id="third">
    <div data-role="header">
      <h1>Test</h1>
    </div>
    <div data-role="content">
      <input type="button" id="pageBtn" value="Go to page">
    </div>
  </div>
  <script>
    $("#pageBtn").click(function() {
      $.mobile.changePage("test2.html", {transition:"flip"});
    });
  </script>
```

The page simply contains one button. At the bottom of the file is a jQuery event listener for that button. When clicked, `$.mobile.changePage` is used to load `test2.html` while making use of the flip transition

Path and URL-related utilities

These utilities are related to the current location, URL, or path of the application:

- `$.mobile.path.isAbsoluteUrl` and `$.mobile.path.isRelativeUrl`: These two functions look at a URL and allow you to check if they are either a complete, absolute, or a relative URL.

- `$.mobile.path.get()`: This returns the "directory" portion of a URL. So given a URL in the form of `http://www.raymondcamden.com/demos/foo.html`, it would return `http://www.raymondcamden.com/demos/`.

- `$.mobile.path.makeUrlAbsolute(relative url, absolute url):`
A slightly different form of the previous function, this utility works
with absolute URLs instead.

Code 9-6 is a **tester** application. It contains the form fields allowing you to test all
of the methods previously discussed. First, let's just look at the UI controls used for
the application:

```
Code 9-6: test4.html
<div data-role="page" id="third">

    <div data-role="header">
       <h1>Test</h1>
    </div>

    <div data-role="content">

        <form>

        <div data-role="fieldcontain">
            <label for="isabsurl">Is Absolute URL?</label>
            <input type="text" name="isabsurl" id="isabsurl" value=""
/>
            <div id="isabsurlresult"></div>
        </div>

        <div data-role="fieldcontain">
            <label for="isrelurl">Is Relative URL?</label>
            <input type="text" name="isrelurl" id="isrelurl" value=""
/>
            <div id="isrelurlresult"></div>
        </div>

        <div data-role="fieldcontain">
            <label for="makeurl">Make URL Absolute</label>
            <input type="text" name="makeurl" id="makeurl" value=""
placeholder="Relative URL" />
            <input type="text" name="makeurl2" id="makeurl2" value=""
placeholder="Absolute URL" />
            <div id="makeurlresult"></div>
        </div>

        <div data-role="fieldcontain">
            <label for="pathget">Path Get</label>
            <input type="text" name="pathget" id="pathget" value=""  />
```

```
        <div id="pathgetresult"></div>
    </div>

    </form>
  </div>

</div>
```

This creates the following form:

Now let's look at the code:

```
Code 9-7: test4.html (continued)
<script>
$("#isabsurl").keyup(function() {
    var thisVal = $(this).val();
    var isAbsUrl = $.mobile.path.isAbsoluteUrl(thisVal);
    $("#isabsurlresult").text(isAbsUrl);
});

$("#isrelurl").keyup(function() {
```

```
      var thisVal = $(this).val();
      var isRelUrl = $.mobile.path.isRelativeUrl(thisVal);
      $("#isrelurlresult").text(isRelUrl);
    });

    $("#makeurl,#makeurl2").keyup(function() {
      var urlVal1 = $("#makeurl").val();
      var urlVal2 = $("#makeurl2").val();
      var makeUrlResult = $.mobile.path.makeUrlAbsolute(urlVal1,urlVal2);
      $("#makeurlresult").text(makeUrlResult);
    });

    $("#pathget").keyup(function() {
      var thisVal = $(this).val();
      var path = $.mobile.path.get(thisVal);
      $("#pathgetresult").html(path);
    });

</script>
```

The previous two code listings are a bit long, but it's really pretty simple.
Each fieldcontain block consists of one particular test of the path methods
and utilities. In the bottom-half of the template, you can see we've made use of
keyup event listeners to monitor changes to these fields and run each test. You
can use this template to see how these methods react based on different inputs.

jQuery Mobile widget and form utilities

We've mentioned numerous times how jQuery Mobile automatically updates various
items and supports things such as lists and collapsible content. One of the things you
may run into, however, is trying to get jQuery Mobile to work with content loaded
after the page is rendered. So, for example, imagine a list view that has data added to
it via some JavaScript code. Code 9-8 demonstrates a simple example of this. It has
a listview with a few items in it, but also a form by which a person could add
new entries:

Code 9-8: test5.html

```
    <div data-role="page" id="third">
      <div data-role="header">
        <h1>List Updates</h1>
      </div>
      <div data-role="content">
        <ul id="theList" data-role="listview" data-inset="true">
```

```
      <li>Initial</li>
      <li>Item</li>
    </ul>
    <form>
      <div data-role="fieldcontain">
        <label for="additem">New Item</label>
        <input type="text" name="additem" id="additem"
          value=""  />
      </div>
      <input type="button" id="testBtn" value="Add It">
    </form>
  </div>
</div>
<script>
  $("#testBtn").click(function() {
    var itemToAdd = $.trim($("#additem").val());
    if(itemToAdd == "") return;
    $("#theList").append("<li>"+itemToAdd+"</li>");
  });
</script>
```

When initially loaded, notice that everything seems fine. However, the following screenshot shows what happens when an item is added to the end of the list:

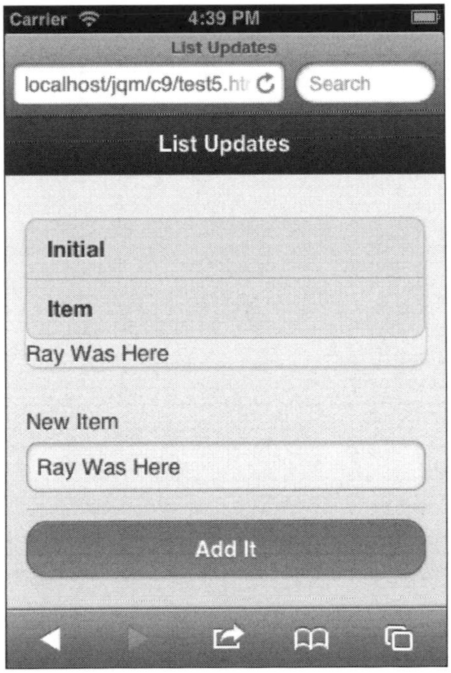

As you can see, the new item was indeed added to the end of the list, but it wasn't drawn correctly. This brings up a critical point; jQuery Mobile parses your code for data attributes and checks for form fields once. After it has done so, it considers its work done. Luckily, there is a standard way for these UI items to be updated. For our `listview` it is a simple matter of calling the `listview` method on the list itself. The `listview` method can be used to turn a new list into a `listview`, or to refresh an existing `listview`. To refresh our `listview`, we'd simply modify the code, as shown in the following code snippet:

```
<script>
  $("#testBtn").click(function() {
    var itemToAdd = $.trim($("#additem").val());
    if(itemToAdd == "") return;
    $("#theList").append("<li>"+itemToAdd+"</li>");
    $("#theList").listview("refresh");
  });
</script>
```

You can find the previous code snippet in `test6.html`. The following screenshot shows how the application handles the new item:

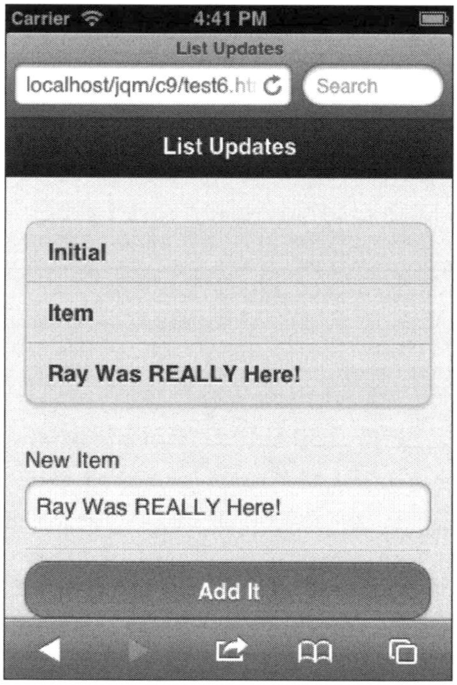

That `listview` method could also be used for completely new lists. Consider the following code snippet Code 9-9:

```
Code 9-9: test7.html
<div data-role="page" id="third">
    <div data-role="header">
      <h1>List Updates</h1>
    </div>
    <div data-role="content" id="contentDiv">
      <input type="button" id="testBtn" value="Add A List">
    </div>
</div>
<script>
  $("#testBtn").click(function() {
    $("#contentDiv").append("<ul data-role='listview' data-
      inset='true' id='theList'><li>Item One</li><li>Item
      Two</li></ul>");
  $("#theList").listview();
  });
</script>
```

In this example, a completely new list is appended to the `div` tag. Notice that we still include the proper `data-role`. But, this by itself, is not enough. We follow up the HTML insertion with a call to the `listview` method to enhance the list just added.

Similar APIs exist for other fields. For example, new buttons added to a page can be enhanced by calling the `button()` method on them. In general, assume any changes to enhanced controls will need to be updated via their respective JavaScript APIs.

Summary

In this chapter, we (finally!) broke out some JavaScript. We looked at how you can configure various jQuery Mobile settings, what utilities exist, and how to handle post-rendered updates to enhanced controls.

In the next chapter, we'll continue working with JavaScript and look at the various events your code can listen to.

10
Working with Events

In this chapter, we will look at how events work in jQuery Mobile. While developers obviously have access to regular events (button clicks, and so on), jQuery Mobile also exposes its own events for developers to use.

In this chapter, we will do the following:

- Discuss touch, swipe, scroll, and other physical events
- Discuss page events

Working with physical events

For the first part of this chapter, we will focus on the "physical" events, or events related to touch and other actions done with a device.

For those of you who have been testing jQuery Mobile using a regular browser, please note that some of the following examples will not work properly on a desktop browser. If you wish, you can download and install emulators for various mobile phone types. For example, Android has an SDK that supports creating virtual mobile devices. Apple also has a way to simulate an iOS device. Setting up and installing these emulators are beyond the scope of this chapter, but it is certainly an option. Of course, you can also use a real hardware device as well.

Chrome users can go into their Dev Tools settings and emulate the touch events. This will be useful for some of the examples in this chapter.

The physical events include the following:

- `tap` and `taphold`: `tap` represents what it sounds like—a quick physical touch on the web page. `taphold` is a longer touch. Many applications will make use of two separate actions, one for `tap` and one for `taphold`.

- `swipe`, `swipeleft`, and `swiperight`: These represent swipes or a finger movement across most of the devices. The `swipe` event is a generic one, whereas `swipeleft` and `swiperight` represent a swipe in a specific direction. There is no support for a swipe up or down event.

- `scrollstart` and `scrollstop`: They respectively handle the beginning and end of scrolling a page.

- `orientationchange`: This is fired when the device's orientation changes.

- `vclick`, `vmousedown`, `vmouseup`, `vmousemove`, `vmousecancel`, and `vmouseover`: All of these are "virtual" events meant to abstract away checking for either touch or mouse-click events. As these are mainly just aliases for click and touch events, they will not be demonstrated.

Now that we've listed the basic physical events, let's start looking at a few examples. `Code 10-1` demonstrates a simple example of the `tap` and `taphold` events:

```
Code 10-1: test1.html
<div data-role="page" id="first">
    <div data-role="header">
      <h1>Tap Tests</h1>
    </div>
    <div data-role="content">
      <p>
        Tap anywhere on the page...
      </p>
      <p id="status"></p>
    </div>
</div>
<script>
  $("body").bind("tap", function(e) {
    $("#status").text("You just did a tap event!");
  });
  $("body").bind("taphold", function(e) {
    $("#status").text("You just did a tap hold event!");
  });
</script>
```

This template is rather simple. The page has some explanatory text asking the user to tap on it. Beneath it is an empty paragraph. Note though the two binds at the end of the document; one listens for `tap` while the other listens for `taphold`. The user can do either action and a different status message is displayed. While rather simple, this gives you a good idea of how you could respond differently based on how long the user holds their finger down. Note that a `taphold` event *also* fires a tap event, specifically when the user lifts their finger off the device. You will need to handle this behavior if you intend to use `taphold` events.

Now let's look at `Code 10-2`, which is an example of swipe events:

```
Code 10-2: test2.html
<div data-role="page" id="first">
    <div data-role="header">
      <h1>Swipe Tests</h1>
    </div>
    <div data-role="content">
      <p>
        Swipe anywhere on the page...
      </p>
      <p id="status"></p>
    </div>
</div>
<script>
  $("body").bind("swipe", function(e) {
    $("#status").append("You just did a swipe event!<br/>");
  });
  $("body").bind("swipeleft", function(e) {
    $("#status").append("You just did a swipe left event!<br/>");
  });
  $("body").bind("swiperight", function(e) {
    $("#status").append("You just did a swipe right
      event!<br/>");
  });
</script>
```

This example is pretty similar to the previous one, except now our event handlers listen for `swipe`, `swipeleft`, and `swiperight`. One important difference is that we append to the `status` div instead of simply setting it. Why? A `swiperight` or `swipeleft` event is automatically a swipe event. If we simply set the text in the paragraph, one would wipe out the other.

The following screenshot shows how the device looks after a few swipes:

How about a more complex example? Consider the following code snippet, Code 10-3:

```
Code 10-3: test3.html
<div data-role="page" id="first">
    <div data-role="header">
      <h1>First</h1>
    </div>
    <div data-role="content">
      <p>
        Swipe to navigate
      </p>
    </div>
  </div>
  <div data-role="page" id="second">
    <div data-role="header">
      <h1>Second</h1>
    </div>
    <div data-role="content">
      <p>
        Swipe to the right...
      </p>
    </div>
  </div>
```

```
<script>
  $("body").bind("swipeleft swiperight", function(e) {
    var page = $.mobile.activePage[0];
    var dir = e.type;
    if(page.id == "first" && dir == "swipeleft")
      $.mobile.changePage("#second");
    if(page.id == "second" && dir == "swiperight")
      $.mobile.changePage("#first");
  });
</script>
```

In this example, we've got a file that includes two separate pages, one with the ID first and the other with the ID second. Notice that we have no links. So how do we navigate? With swipes! Our event handler is now listening for both swipeleft and swiperight. We first grab the active page using $.mobile.activePage, as described in *Chapter 9, jQuery Mobile Configuration, Utilities, and JavaScript Methods* on methods and utilities. The [0] at the end refers to the fact that the value is actually a jQuery Selector. Using [0] grabs the actual DOM item. The event type will be either swipeleft or swiperight. Once we know that, we can actively move the user around depending on what page they are currently on and in what direction they swiped.

Now let's look at scrolling. You can detect when a scroll starts and when one ends. Code 10-4 is another simple example of this in action:

```
Code 10-4: test4.html
<div data-role="page" id="first">
    <div data-role="header">
      <h1>Scroll Tests</h1>
    </div>
    <div data-role="content">
      <p>
        Scroll please....<br/>
        <br/>
        <br/>
        (Many <br/> tags removed to save space!)
        <br/>
        <br/>
      </p>
      <p id="status"></p>
    </div>
</div>
<script>
  $("body").bind("scrollstart", function(e) {
    $("#status").append("Start<br/>");
  });
  $("body").bind("scrollstop", function(e) {
    $("#status").append("Done!<br/>");
  });
</script>
```

This template is pretty similar to `test1.html`, the tap tester, except now we've listened to `scrollstart` and `scrollstop`. Also note the list of `<br/>` tags. In the real source file, there are many of these. This will ensure that the page is actually scrollable when you test. When the scrolling will start and end, we simply append to another `status` `div`. (Please note that currently DOM manipulation is listed as being buggy when listening to `scrollstart`. The previous example may not work in iOS, but works fine on Android.)

Now let's look at orientation events in `Code 10-5`:

```
Code 10-5: test5.html
<div data-role="page" id="first">
    <div data-role="header">
      <h1>Orientation Tests</h1>
    </div>
    <div data-role="content">
      <p>
        Tilt this sideways!
      </p>
      <p id="status"></p>
    </div>
</div>
<script>
  $(window).bind("orientationchange", function(e,type) {
    $("#status").html("Orientation changed to "+e.orientation);
  });
</script>
```

The critical part of the previous code snippet is the JavaScript at the end, specifically the event listener for changing orientation. This is not actually a jQuery Mobile supported event but something supported by the browser itself. Once the event listener is attached, you can do whatever you wish based on the orientation of the device. The following screenshot is a demonstration:

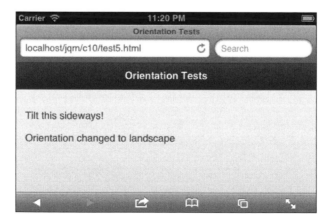

Handling page events

Now that we've discussed physical type events, it's time to turn our attention to page events. Remember that jQuery Mobile has its own concept of pages. In order to give developers even more control over how pages work within jQuery Mobile, numerous page events are supported. Not all will necessarily be useful in your day-to-day development. In general, page events can be split into the following categories:

- **load**: These are events related to the loading of a page. They are `pagebeforeload`, `pageload`, and `pageloadfailed`. `pagebeforeload` is fired prior to a page being requested. Your code can either approve or deny this request based on whatever logic may make sense. If a page is loaded, `pageload` is fired. Conversely, `pageloadfailed` will be fired on any load that does not complete.

- **change**: These events are related to the change from one page to another. They are: `pagebeforechange`, `pagechange`, and `pagechangefailed`. As before, the `pagebeforechange` function acts as a way to programmatically decline the event. If done, the `pagechangefailed` event is fired. `pagebeforechange` is fired *before* the `pagebeforeload` event. `pagechange` will fire after the page is displayed.

- **transition**: These events are related to the movement, or transition, from one page to another. They are `pagebeforeshow`, `pageshow`, `pagebeforehide`, and `pagehide`. Both `pagebeforeshow` and `pagebeforehide` run prior to their related events but unlike `pagebeforeload` and `pagebeforechange`, they can't actually prevent the next event.

- **init**: As it has been shown many times in this book, jQuery Mobile performs multiple updates to basic HTML to optimize it for mobile displays. These are initialization-related events. The events you can listen to are `pagebeforecreate`, `pagecreate`, and `pageinit`. `pagebeforecreate` fires before any of the automatic updates are fired on your controls. This allows you to manipulate your HTML via JavaScript beforehand. `pagecreate` is fired after page content exists in the DOM, but still before the layout has been updated by jQuery Mobile. The official docs recommend this as the place to do any custom widget handling. Finally, `pageinit` will run after the initialization has been completed.

- **remove**: There is one event for this category called `pageremove`. This event is fired before jQuery Mobile removes an inactive page from the DOM. You can listen to this event to prevent the framework from removing the page.

- **layout**: The final category is related to layout and has one event called `updatelayout`. This is typically fired by other layout changes as a way to let the page know it needs to update itself.

That's quite a lot! A simple way to look at these events in action would be to simply listen to all of them. In Code 10-6, we have a simple example of this in action:

```
Code 10-6: test_page.html
<div data-role="page" id="first">
    <div data-role="header">
      <h1>Page Event Tests</h1>
    </div>
    <div data-role="content">
      <p>
        <a href="#page2" data-role="button">Go to Page 2</a>
        <a href="test_pagea.html" data-role="button">
          Go to Page 3</a>
        <a href="test_pageb.html" data-role="button">
          Go to Page 4</a>
        <a href="test_pageDOESNTEXIST.html" data-role="button">
          Go to Page Failed</a>
      </p>
    </div>
</div>
<div data-role="page" id="page2">
    <div data-role="header">
      <h1>Page Event Tests</h1>
    </div>
    <div data-role="content">
      <p>
        <a href="#first" data-role="button">Go to Page 1</a>
        <a href="test_pagea.html" data-role="button">
          Go to Page 3</a>
        <a href="test_pageb.html" data-role="button">
          Go to Page 4</a>
      </p>
    </div>
</div>
<script>
  $(document).bind("pagebeforeload pageload pageloadfailed
    pagebeforechange pagechange pagechangefailed pagebeforeshow
    pagebeforehide pageshow pagehide pagebeforecreate pagecreate
    pageinit pageremove updatelayout", function(e) {
    console.log(e.type);
  });
</script>
```

This template is part of a four-page, three-file simple application that has buttons linking to each of the other pages. The other pages may be found in the ZIP file you downloaded. In order to test this application, you *should* use a desktop browser with console support. That's any version of Chrome, recent Firefox browsers (or Firefox with Firebug), and the latest Internet Explorer. A full explanation of the browser console wouldn't fit in this chapter, but you can think of it as a hidden-away debugging log useful for recording events and other messages. In this case, we've told jQuery to listen for all of our jQuery Mobile page events. We then log the specific event type to the console. After clicking around a bit, the following screenshot shows how the console log looks in a Chrome browser:

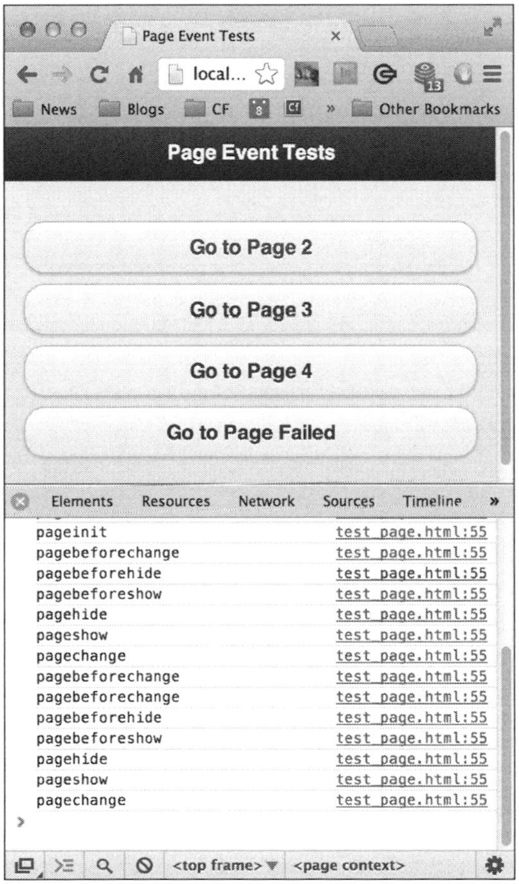

Opening the console in Chrome is simple. Click on the "three line" icon in the upper-right corner of the browser. Select **Tools** and then **JavaScript Console**. Open the console up before testing these files yourself and you can monitor the page events as they happen in real time.

What about $(document).ready?

If you are a jQuery user, you may be curious how $(document).ready comes into play with a jQuery Mobile site. Almost all jQuery applications use $(document).ready for initialization and other important setup operations. However, in a jQuery Mobile application, this will not work well. Because Ajax is used to load pages, $(document).ready is only really effective for the *first* page. Therefore, the pageInit event should be used in cases where you would have used $(document).ready in the past.

Creating a real example

So what about a real example? Our next set of code is going to demonstrate how to create a simple, but dynamic, jQuery Mobile website. The content will be loaded via Ajax. Normally, this would be dynamic data, but for our purposes, we'll use simple static files of JSON data. **JavaScript Object Notation (JSON)** is a way to represent complex data as simple strings. Code 10-7 is the application's home page:

```
Code 10-7: test_dyn.html
<div data-role="page" id="homepage">
    <div data-role="header">
      <h1>Dynamic Pages</h1>
    </div>
    <div data-role="content">
      <ul id="peopleList" data-role="listview"
        data-inset="true"></ul>
    </div>
</div>
<script>
  $("#homepage").bind("pagebeforecreate", function(e) {
  //load in our people
  $.get("people.json", {}, function(res,code) {
    var s = "";
    for (var i = 0; i < res.length; i++) {
    s+="<li><a href='test_people.html
      ?id="+res[i].id+"'>"+res[i].name+"</a></li>";
  }
    $("#peopleList").html(s).listview("refresh");
  }, "json");
});
```

```
$(document).on("pagebeforeshow", "#personpage", function(e) {
    var thisPage = $(this);
    var thisUrl = thisPage.data("url");
    var thisId = thisUrl.split("=")[1];
    $.get("person"+thisId+".json", {}, function(res, code) {
    $("h1",thisPage).text(res.name);
    s = "<p>"+res.name +" is a "+res.gender+" and
        likes "+res.hobbies+"</p>";
    $("#contentArea", thisPage).html(s);
    }, "json");
    });
</script>
```

The first thing you may notice about this jQuery Mobile page is that there isn't any actual content. Not within the jQuery Mobile page's content block at least. There's a `listview` but no actual content. So where's the content going to come from? At the bottom of the page you can see two event listeners. For now, let's just focus on the first one.

The code here binds to the `pagebeforecreate` event that jQuery Mobile fires for pages. We've told jQuery Mobile to run this event before it creates the page. This event will run once and only once. Within this event we use the jQuery `get` feature to do an Ajax request to the `people.json` file. This file is simply an array of names in JSON format:

```
[{"id":1,"name":"Raymond Camden"},{"id":2,"name":"Todd
    Sharp"},{"id":3,"name":"Scott Stroz"},{"id":4,"name":"Dave
    Ferguson"},{"id":5,"name":"Adam Lehman"}]
```

Each name has both an ID and the actual name value. When loaded by jQuery, this is turned into an actual array of simple objects. Looking back at the event handler, you can see that we simply loop over this array and create a string representing a set of `li` tags. Note that each one has a link to `test_people.html` as well as a dynamic name. Also note the links themselves are dynamic. They include each person's ID value as retrieved from the JSON string.

It was mentioned earlier, but take note of the call to `listview("refresh")`:

```
$("#peopleList").html(s).listview("refresh");
```

Without the `listview("refresh")` portion, the items we added to the listview would not be styled correctly.

Next, let's take a quick look at `test_people.html`:

```
Code 10-8: test_people.html
<div data-role="page" id="personpage">
    <div data-role="header">
      <h1></h1>
    </div>
    <div data-role="content" id="contentArea">
  </div>
</div>
```

As with our last page, this one is pretty devoid of content. Note that both the header and the content area are blank. But, if you remember the second event handler in `test_dyn.html`, we have support to load the content here. This time we used the `pagebeforeshow` event. Why? We want to run this code before every display of the page. We need to know what particular person to load. If you remember, the ID of the person was passed in the URL. We can fetch that via a data property, such as `url`, that exists on the page object. This returns the complete URL, but all we care about is the end of it — our ID. So we split the string and grab the last value. Once we have, we can then load in a particular JSON file for each person. The form of this filename is `personX.json`, where X is the number 1 through 5. The following line of code is one example:

```
{"name":"Raymond Camden","gender":"male","hobbies":"Star Wars"}
```

Obviously, a real person object would have a bit more data. Once we fetch this string, we can then parse it and lay out the result on the page itself:

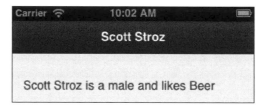

Summary

In this chapter, we looked into events that jQuery Mobile application can listen and respond to. These events include physical types (scrolling, orientation, and touching) and page-based ones as well.

In the next chapter, we'll look at how jQuery Mobile sites are themed — both out-of-the-box themes and custom ones as well.

11
Enhancing jQuery Mobile

In this chapter we'll be learning about how to enhance jQuery Mobile, and how to make our mobile application really stand out from the pack by creating themes and icons to improve the look and functionality of our app. We will cover the following aspects:

- Learn about the building blocks of jQuery Mobile
- Create our own jQuery Mobile theme using ThemeRoller
- Design and implement custom icons for our application

What's possible?

The first reaction many developers have when first using jQuery Mobile is to awe at how easy it is to implement a rich, compelling mobile website for their users. The ease with which it converts plain HTML to beautiful, usable buttons, listviews, and form elements is a dream. The jQuery Mobile team shipped five well-designed and attractive themes, and 20 commonly used icons along with the rest of the package. They even built a tool which allows developers to build out their own themes: ThemeRoller (`http://jquerymobile.com/themeroller/`).

After working with jQuery Mobile for a while, some developers might be asking, "What else can I do with this?" Just like muscle cars from the 60s and 70s; it wasn't enough that they were already awesome; the tweakers and the gearheads wanted to do more. If you identify with that mentality, then this chapter is for you.

The wonderful thing about jQuery Mobile is that it's just HTML, and for that reason we can do great things with very little effort. In this chapter we'll be creating our own theme from scratch using ThemeRoller for jQuery Mobile. We'll be designing buttons from scratch and writing the CSS code needed to implement both low and high resolution versions. We'll also be looking at how we can expand on the styles and classes already available in jQuery Mobile, and make something different and unique. Let's get started, shall we?

The visual building blocks of jQuery Mobile

As you've already seen, jQuery Mobile is very user-friendly and pleasing to the eye. It uses rounded corners, subtle gradients, and drop shadows to make elements stand out from their surroundings, and other tricks that graphic designers have been using for years in print. But on the web, these effects were only possible with the use of images, or complicated and poorly supported plugins and applets.

With the advent of Web 2.0 and CSS 3, all of these options have been made available to us, the layman web developer. Just remember that with great power comes great responsibility. jQuery Mobile operates on the principle of progressive enhancement. A tricky phrase, but it just means that you should develop for the lowest common denominator, and offer enhancements for browsers that understand them.

Lucky for us, these stylistic additions are almost purely cosmetic. If a browser doesn't understand the border-radius declaration then it simply displays squared off corners. The same holds true for gradients and shadows. While jQuery Mobile adds these effects to your application out of the box, it's worthwhile knowing how to add them on your own.

Border-radius

Rounded corners can be one of the most elegant and appealing effects, and are also the simplest to add. There are a few caveats that developers need to know about this and the other effects. While there is a spec for border-radius as recommended by the W3C, it turns out that each of the primary browser manufacturers support it in slightly different ways. The end result is the same, but the path to it varies. Let's take a look at the most basic border-radius declaration, and its result:

```
#rounded {
  border-radius: 10 px;
}
```

This box has rounded corners.

You also have the option of rounding only certain corners, as well as tweaking the values so that the corner isn't a perfect quarter-circle. Let's look at a few more examples:

```
#topLeftBottomRight {
  border-radius: 15px 0 15px 0;
}
```

```
#topLeft {
  border-top-left-radius: 100px 40px;
}
```

Sadly, it's not quite as easy as this, just yet. Because each browser vendor has their own unique rendering for this effect; software developers like Google or Mozilla have taken to creating their own versions, called vendor prefixes. For the preceding style declarations to have the widest range of coverage, you'd have to add the following lines:

```
#rounded {
-webkit-border-radius: 10 px;
-moz-border-radius: 10 px;
border-radius: 10 px;
}

#topLeftBottomRight {
-webkit-border-top-left-radius: 15px;
-webkit-border-bottom-right-radius: 15px;
-moz-border-radius-topleft: 15px;
-moz-border-radius-bottomright: 15px;
border-top-left-radius: 15px;
border-bottom-right-radius: 15px;
/* Mozilla and webkit prefixes require you to define each corner
individually when setting different values */
}

#bottomLeft {
```

```
-webkit-border-top-left-radius: 100px 40px;
-moz-border-radius-topleft: 100px 40px;
border-top-left-radius: 100px 40px;
}
```

Applying drop shadows

Drop shadows in CSS take one of two forms: text shadows (applied to text) and box shadows (applied to everything else). Like `border-radius`, drop shadows are fairly straightforward if you're looking at the W3C spec.

Using text-shadow

Let's look at `text-shadow` first:

```
p {
  text-shadow: 2px 2px 2px #000000; /* horizontal, vertical, blur,
    color */
}
```

The text in this paragraph will have a drop shadow that is offset 2 pixels to the right, 2 pixels to the bottom, is blurred by 2 pixels, and whose color is black.

This property also supports multiple shadows by adding additional declarations in a comma-separated list:

```
p {
  text-shadow: 0px 0 px 4px white,
  0 px -5px 4px #ffff33,
  2px -10 px 6px #ffdd33,
  -2px -15px 11px #ff8800,
  2px -25px 18px #ff2200
}
```

How about a cheesy fire effect!

Unlike the `border-radius` property, the `text-shadow` property doesn't require vendor prefixes. That doesn't mean that all browsers support it; it simply means that browsers that do support this property will display it as intended, while browsers that do not, cannot simply see anything.

Using box-shadow

Box-shadow follows a very similar model to text-shadow with one addition, the inset keyword that allows for inner shadowing. Let's get to the examples. First up, standard outer shadows:

```
#A {
  -moz-box-shadow: -5px -5px #888888;
  -webkit-box-shadow: -5px -5px #888888;
  box-shadow: -5px -5px #888888; /* horizontal, vertical, color */
}

#B {
  -moz-box-shadow: -5px -5px 5px #888888;
  -webkit-box-shadow: -5px -5px 5px #888888;
  box-shadow: -5px -5px 5px #888888; /* horizontal, vertical,
    blur, color */
}

#C {
  -moz-box-shadow: 0 0 5px 5px #888888;
  -webkit-box-shadow: 0 0 5px 5px #888888;
  box-shadow: 0 0 5px 5px #888888; /* horizontal, vertical, blur,
    spread, color */
}
```

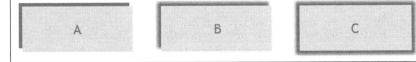

And now, check out these inner shadows. Snazzy eh?

```
#D {
  -moz-box-shadow: inset -5px -5px #888888;
  -webkit-box-shadow: inset -5px -5px #888888;
  box-shadow: inset -5px -5px #888;}

#E {
  -moz-box-shadow: inset -5px -5px 5px #888888;
  -webkit-box-shadow: inset -5px -5px 5px #888888;
  box-shadow: inset 0px 0px 10 px 20px #888888;
}

#F {
```

```
  -moz-box-shadow: inset -5px -5px 0 5px #888888;
  -webkit-box-shadow: inset -5px -5px 0 5px #888888;
  box-shadow: inset 0 0 5px 5px #888888;
}
```

It's worth mentioning that both box-shadow and text-shadow can have their colors set with the less commonly used rgb and rgba declarations. This allows developers to set colors using the more familiar convention of RGB values. The rgba declaration also allows the setting of color opacity from 0 to 1. The code for that is a simple change:

```
#opacity {
  box-shadow: inset 0 0 5px 5px rgb(0,0,0); /* black */
}
#transparent {
  box-shadow: inset 0 0 5px 5px rgba(0,0,0,.5); /* black with 50%
    opacity */
}
```

CSS gradients

CSS gradients are a great way to add beauty and impact to your website. The options include linear gradients (right to left, top to bottom, and so on), and radial gradients (from center outwards). By default, gradients consist of a start color and an end color. CSS gradients may also include additional tones using color stops.

It should be noted, however, that support for CSS gradients in older browsers isn't perfect, specifically Internet Explorer. The good news is that there are ways to address IE that can allow developers to reliably use gradients in their development. The bad news is that the code for that support is robust. Let's take a look at the simplest possible gradient declaration.

```
div {
  width: 500px;
  height: 100px;
  background: linear-gradient(left,  #ffffff 0%,#000000 100%);
}
```

Gradient declarations can be quite complex, so let's break it down with an infographic:

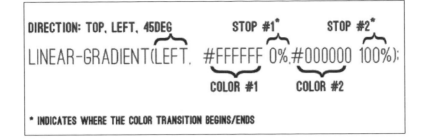

Now here's the kicker. At the time of this writing there were no browsers that supported the W3C spec using the actual property. Let's take a look at the code to support multiple browsers, and you'll love jQuery Mobile even more than you already do.

```
div {
  width: 500px;
  height: 100px;
  border: 1px solid #000000;
  /* Old browsers */
  background: #ffffff;
  /* FF3.6ı */
  background: -moz-linear-gradient(left,  #ffffff 0%, #000000
    100%);
  /* Chrome10+,Safari5.1+ */
  background: -webkit-linear-gradient(left,  #ffffff 0%,#000000
    100%);
  /* Opera 11.10+ */
  background: -o-linear-gradient(left,  #ffffff 0%,#000000 100%);
  /* IE10+ */
  background: -ms-linear-gradient(left,  #ffffff 0%,#000000 100%);
  /* W3C spec*/
  background: linear-gradient(left,  #ffffff 0%,#000000 100%);
  /* IE6-9 */
  filter: progid:DXImageTransform.Microsoft.gradient(
    startColorstr='#ffffff', endColorstr='#000000',GradientType=1
);
}
```

You can add multiple colors to your gradient by adding additional comma-separated declarations. Consider the following code:

```
div {
  width: 500px;
  height: 100px;
  border: 1px solid #000000;
  /* Old browsers */
  background: #ffffff;
  /* FF3.6+ */
  background: -moz-linear-gradient(left,  #ffffff 0%, #000000 35%,
    #a8a8a8 100%);
  /* Chrome10+,Safari5.1+ */
  background: -webkit-linear-gradient(left,  #ffffff 0%,#000000
    35%,#a8a8a8 100%);
  /* Opera 11.10+ */
  background: -o-linear-gradient(left,  #ffffff 0%,#000000
    35%,#a8a8a8 100%);
  /* IE10+ */
  background: -ms-linear-gradient(left,  #ffffff 0%,#000000
    35%,#a8a8a8 100%);
  /* W3C */
  background: linear-gradient(left,  #ffffff 0%,#000000
    35%,#a8a8a8 100%);
  /* IE6-9 */
  filter: progid:DXImageTransform.Microsoft.gradient(
    startColorstr='#ffffff', endColorstr='#a8a8a8',GradientType=1
    );
}
```

The previous code results in the following gradient:

As you might guess after reading the last few pages, jQuery Mobile does a lot of heavy lifting on your behalf. Not only does it add slick gradients as page backgrounds, but it has to keep track of all of the browser quirks that might prevent that sweet drop shadow from showing up. As we move into the next section, you'll likely be even more impressed with the way it handles themes and color swatches.

The basics of jQuery Mobile theming

Theming in jQuery Mobile is straightforward and simple to use for the developer, but is pretty elaborate behind the scenes. Luckily there will rarely be a time when you need to know everything that's being done for you. However, it's worth a little bit of our time to understand how it works.

Out of the box, jQuery Mobile comes with a theme set comprised of five color swatches, each associated with a letter from A to E. The theme contains a series of base CSS classes that can be applied at will to nearly any element, and contain global settings for width, height, border radius, and shadows. The individual swatches contain specific information about color, fonts, and so on.

Additional swatches can be added to the five original swatches from F to Z, or the original swatches can be replaced or overridden at will. This system allows for a total of 26 distinct swatches, allowing for millions of possible combinations of theme colors, styles, and patterns. You apply a jQuery Mobile theme to the selected element by adding a `data-theme` attribute with the letter of the desired theme.

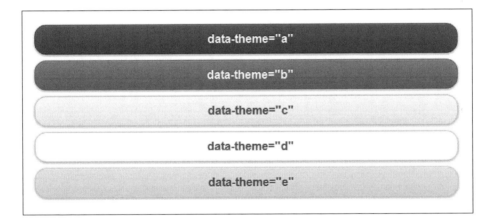

Developers will generally choose to use the `data-theme` attribute method when applying styles, but it's also possible to attach the CSS class names directly to your page elements for more granular control. There are a handful of primary prefixes that allow for this flexibility.

Bars (.ui-bar-?)

The bar prefix is generally applied to headers, footers, and other areas with high importance:

Bar A - .ui-bar-a
Bar B - .ui-bar-b
Bar C - .ui-bar-c
Bar D - .ui-bar-d
Bar E - .ui-bar-e

Content blocks (.ui-body-?)

Content blocks are generally applied to areas where paragraph text is expected to occur. Its color helps to ensure maximum readability with the text color placed against it:

Block A - .ui-body-a
Block B - .ui-body-b
Block C - .ui-body-c
Block D - .ui-body-d
Block E - .ui-body-e

Buttons and listviews (.ui-btn-?)

Buttons and listviews are two of the most important elements in the jQuery Mobile library, and you can rest assured that the team took their time getting these right. The .ui-btn prefix also includes styles for up, down, hover, and active states:

Mixing and matching swatches

One of the nice things about theming in jQuery Mobile is that child elements inherit from their parent unless otherwise specified. This means that if you put a button without its own data-theme attribute inside a header or footer bar, that button will use the same theme as its parent. Wicked eh?

It's also perfectly acceptable and even encouraged to place an element using one swatch as the child of an element using another swatch. This can help the element stand out more, match a different part of the app, or whatever reasoning the developer chooses. It's possible, and what's more, it's easy. Simply place a button (or other element) inside a header bar, and assign it its own `data-theme` attribute:

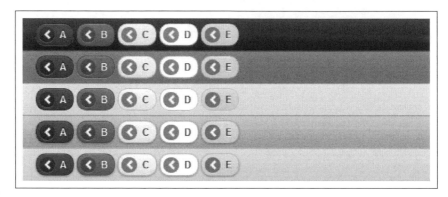

Site-wide active state

jQuery Mobile also applies a global active state for all elements. This active state is used for buttons, form elements, navigation, and anywhere there's a need to indicate that something is currently selected. The only way to change this color value is to set (or override) it via CSS. The CSS class for the active state is, appropriately named, `.ui-btn-active`:

Default icons

Included in the jQuery Mobile set are 20 icons which cover a wide array of needs for developers. The icon set is white on transparent which jQuery Mobile overlays with a semi-transparent black circle to provide contrast against all of the swatches. To add an icon, specify the `data-icon` attribute with the name of the desired icon. In addition to the white icon set, there is also a black on transparent icon set for colors that need higher contrast. You can force any icon to use the dark set by adding the `ui-icon-alt` class to any element that uses an icon.

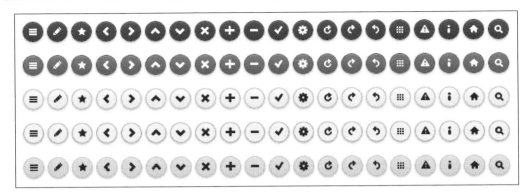

jQuery Mobile also provides the ability to place icons on the top, right, bottom, or left side of a button using the `data-iconpos="[top, right, bottom, left]"` attribute, with left being the default placement. Developers are also able to display an icon alone, without text, by specifying `data-iconpos="notext"`:

Custom icons are also possible and will be covered later in this chapter.

Creating and using a custom theme

We've already discussed how powerful theming is in jQuery Mobile. It makes it trivial to develop a rich mobile website with simple and elegant style. Even more powerful is the ability to create your own library of swatches that can be used to make your application or website truly unique. Developing your own theme can be approached in one of the following two ways:

1. Download and open the existing jQuery Mobile CSS file and edit to your heart's content.
2. Point your web browser to ThemeRoller for jQuery Mobile: http://jquerymobile.com/themeroller/.

We'll be focusing solely on option 2, because let's be honest, why wade through all of that CSS when you can point, click, and drag your way to a new theme full of swatches in like 10 minutes? Let's find out what ThemeRoller is all about.

What's ThemeRoller?

ThemeRoller for jQuery Mobile is an extension of a web-based app that was written for the jQuery UI project. It allows users to quickly assemble a theme full of swatches in minutes by pointing, clicking, and dragging. It features an interactive preview so that you can immediately see how your changes affect your theme. It also has a built-in inspector tool that helps you dig into the minute details (should you want them). It also integrates with Adobe® Kuler®, a color management tool. You can download your theme after you're done, you can share it with others via a custom URL, and you can reimport past themes for last-minute tweaking. It's a powerful tool and is a perfect complement to jQuery Mobile.

One of the hallmarks of the five default swatches is that the jQuery Mobile team spent quite a bit of time working on readability and usability. The swatches range from highest contrast (A), to lowest contrast (E). Within a single theme, the areas that have the most contrast are the areas most prominent on the page. This includes the header (and listview headers), and buttons. When creating your own theme, it's a good idea to keep this in mind. We always want to focus on usability within our app, right? What good is a slick app if no one can read it because of poor color choices?

Using ThemeRoller

The first thing you'll see when you load up ThemeRoller is a slick-looking splash screen.

Following the splash screen is a helpful "Getting Started" screen (the "Getting Started" screen has some helpful tips, so make sure to glance at it before clicking the **Get Rolling** button).

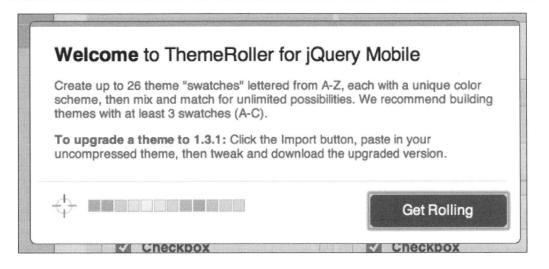

After all of the splash screens are out of the way, you'll be presented with the primary interface:

ThemeRoller is broken up into four main areas: **Preview**, **Color**, **Inspector**, and **Tools**. Each of these contains important functionality that we need to review. We'll start with the Preview section.

Preview

Unless you're loading an existing theme, the preview area will present three complete, identical, and interactive jQuery Mobile pages packed with widgets of all sorts:

Move your mouse over them and you'll see that each page is functional. The header on each page contains a letter indicating which swatch controls its appearance.

Colors

At the top of the page, you'll see a series of color chips along with two slider controls and a toggle button. Further to the right, you'll see another 20 blank chips. These will contain your recently used colors and will be empty until you've selected a color.

Below the color chips are two sliders labeled **Lightness** and **Saturation**. The **Lightness** slider adjusts the light and dark tones of the series of colors swatches, while the **Saturation** slider makes the colors more or less vibrant. Taken together, a user should be able to approximate nearly any color they choose. To use colors from Kuler®, click on the text link marked **Adobe Kuler swatches**.

Each of the color chips can be dragged onto any element within the preview area. This makes development of a swatch set extremely easy. Note that many of the jQuery Mobile styles overlap; for example, the header bar at the top of the page receives the same style as the header of the listview. Adjust the colors as desired then drag each chip onto an element on the page. Remember that each individual page has its own swatch so be careful about how you choose to mix colors.

Inspector

On the far left of the interface is the **Inspector** panel that allows you to exercise fine grain control over your theme:

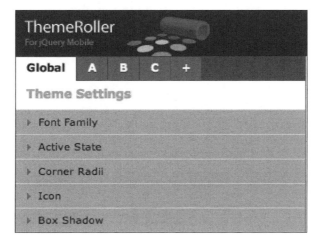

The bottom section contains a series of tabs labeled **Global**, **A**, **B**, **C**, and **+**. Each tab contains an accordion panel with all of the values for an individual swatch, except for the **Global** tab that applies to all of the swatches.

Click on the **Global** tab, then click on **Active State**, and the accordion panel will expand to show settings for the active state for your entire theme. The options include text color, text shadow, background, and border. Changing a value in the **Global** tab causes every current (and future) swatch to reflect the new setting.

Additional swatches can be added to your theme in two ways. Clicking the **+** tab at the top of the **Inspector** adds a new swatch to the last place in your theme. You can also add a new swatch by clicking the **Add Swatch** button located at the bottom of the preview area. Swatches can be deleted by selecting the tab with the swatch you want to remove, then clicking on the **Delete** link located to the right of the swatch name. Note that deleting a swatch from the top of the stack will cause the remaining swatches to be renamed.

Tools

At the very top of the page are a series of buttons. These buttons allow you to perform a variety of tasks which we'll cover in a moment, but first, let's take a closer look at the buttons themselves:

You'll notice the following buttons: a switch allowing you to change between current and previous versions, undo/redo, and a toggle button for the **Inspector**. Setting this toggle to ON allows you to inspect any of the widgets in the preview area. Hovering over a widget highlights that element with a blue box. Clicking on the element will cause the accordion menu in the **Inspector** area to expand to display settings specific to that element.

There are four additional buttons which allow you to download your theme, import, or upgrade a previously created theme, share your theme with others, and a help button.

Creating a theme for Notekeeper

Now that we're familiar with the ThemeRoller interface, how about we go ahead and create our first theme? Rather than build one in abstract let's create one that we'll actually use for the Notekeeper app we built earlier. Let's start simple by modifying one of the existing themes that ship with jQuery Mobile. The team was kind enough to let users import the default themes as a starting place for new themes, so that's where we'll head first. Click on the **Import** button at the top left of the window and you'll get a box allowing you to paste in the contents of an existing theme:

Import the default theme by clicking on the link in the top-right corner, appropriately titled **Import Default Theme**. After the text area fills with CSS, click on **Import**. The preview area will reload and display swatches A to E.

We'll focus our efforts on changing up the white swatch, D, as it's the closest to what our end goal is. Since we'd rather use swatch A as the name, let's delete the other swatches so that only D is left. Remember that ThemeRoller renames swatches as others are deleted. That means when you delete swatch A, B becomes A, C becomes D, and so on. Keep going until the swatch that was D is now in the A position. Finally, delete swatch B (which was formerly swatch E) so that all we have left is swatch A:

This swatch is nice looking but it's a little bland. Let's inject a little color by changing the header to a nice green. The simplest way to determine what values should be changed for any element is to use the **Inspector**. Toggle the **Inspector** to ON at the top, then click anywhere on the header of theme A. You'll know if you got it right if the A tab is selected on the left, and the **Header/Footer Bar** panel expands:

You can change the color in one of a few ways. You can drag a color chip from the top directly onto the background. You can also drag a color chip onto an input field. Finally, you can manually input the value. Notice that when you click into a field containing a color value, you're provided with a slick color picker. Go ahead and change the values in the input fields in this panel to the values shown in the preceding screenshot.

Looking good, but now the blue from the theme's active state clashes with our green. Using the Inspector tool, click once on the **On** section of the **On/Off** toggle bar. This should cause the **Active State** panel within the global tab to expand. We'll change the blue to a nice warm grey. The **Global** panel should now look like the following screenshot:

There's only one thing that's keeping our new theme from looking its hottest; the blue text link in the paragraph at the top. Going back to our trusty Inspector, lets' click directly on the link that will expand the **Content Body** panel within the A tab. Now, for those already familiar with CSS, you know that you can't simply change the link color without also changing the hover, **visited:hover**, and active states.

The problem is that there are no options to make those changes, but ThemeRoller has you covered. Click on the **+** to the right of the **LINK COLOR** input field to display the rest of the options, then change the colors as shown in the following screenshot:

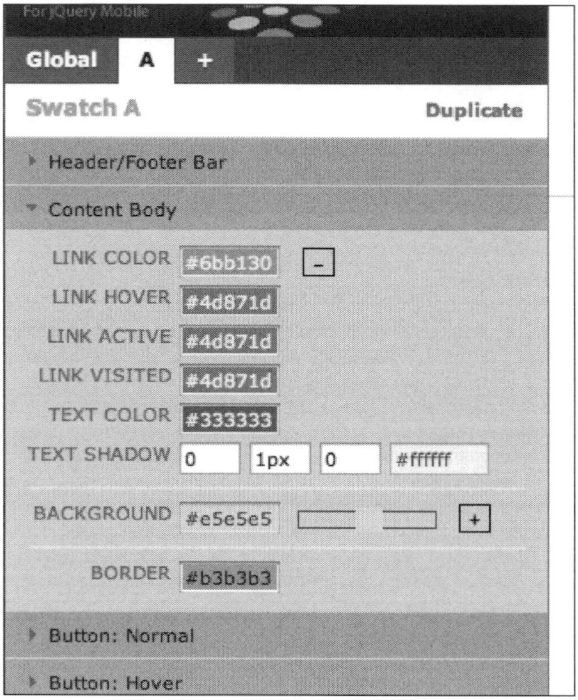

And that's it! Feel free to make additional changes to your theme as you explore the **Inspector** area. Change whatever you like; it's just bits and bytes right now.

Exporting your theme

Before we actually export our theme, there's one thing that must be noted. Remember the splash page with the "helpful" information? It turns out that there's one piece that's not a recommendation, but a requirement.

We recommend building themes with at least 3 swatches (A to C).

For our theme to apply to our Notekeeper app properly, we'll need to duplicate our single swatch (letter A) to swatches B and C. Luckily, this is an easy thing to do. Click on the **A** tab at the top of the **Inspector**, then click the **+** tab twice. What you should see are three identical swatches, and now we're done.

Now that we've finished our theme, we're going to export it for use in our Notekeeper application. This is a straightforward process that begins by clicking on the **Download Theme** button located in the middle of the page at the top of the interface. You'll be presented with a box allowing you to name your theme, some information about how to apply your theme, and a button labeled **Download Zip**. After naming our theme Notekeeper, click on the **Download Zip** button, and you'll receive a tasty little morsel in your downloads folder.

Extract the contents of the ZIP file and you'll see the following directory structure:

- index.html
- themes/
    - Notekeepcr.css (the uncompressed version of your theme)
    - Notekeeper.min.css (the compressed version; can be used in production)
    - images/
    - ajax-loader.gif
    - icons-18-black.png
    - icons-18-white.png
    - icons-36-black.png
    - icons-36-white.png

The HTML file at the top of the tree contains information on how to implement your theme, as well as a few widgets to confirm that the theme works. All of the links are relative in the example file, so you should be able to drag it into any browser window and see the results.

A few notes about the download and implementation of themes are as follows:

- The jQuery team provides the icons for buttons to use in this ZIP file for a reason. The theme requires those images to be relative to the CSS file. This means that unless you're already using the default themes you need to also include the images folder when you upload your theme to your website, or the icons won't show up.
- Hang on to the uncompressed version of your theme. While you don't want to use it in production because of the size, you will need it should you ever wish to edit it within ThemeRoller. ThemeRoller cannot import the minified CSS file.

Creating and using custom icons

We've seen how easy it is to add our own theme to jQuery Mobile using ThemeRoller. Now we're going to add a little more spice to our Notekeeper application by creating a custom icon. The directions in this section will be specific to Photoshop, but any graphics application capable of exporting transparent PNG files should be acceptable.

CSS Sprites

Before we create and use an icon, we should first understand how jQuery Mobile uses icons and applies them. In the theme you just created are several image files (themes/images). Open `icons-18-black.png`, and `icons-36-black.png` in the graphics editor of your choice. Zoom in on both of them to 400 percent or so, and you should see something very similar to the following screenshot:

When opening each of these files, you'll probably notice that each image contains all of the icons. This is because jQuery Mobile takes advantage of a technique called sprite sheets that allows developers to crop a background image by specifying its position within its container, and to hide any other part of the background that would normally display outside of that container. Its primary benefits include the following:

- Reducing the number of requests a browser has to make. Fewer requests generally mean that a page will load faster.

- Centralize image locations. All icons can be found in one location.

Here's a simple illustration of the technique:

A browser always refers to an image from its top left corner. In CSS language that's 0,0. To achieve this affect you set the background image on a container then simply adjust the X and Y coordinates until the image's position matches your design. Then set the overflow of the container to crop, or hide the remainder of the image. Remember that you're moving the image to the left, so you'll use negative numbers for the X position. Using the preceding illustration as a reference, this is what the code to achieve this effect would look like.

```html
<html>
  <head>
    <title></title>
    <style>
      div {
        background: url("icons-36-black.png");
        background-position: -929px 4px;
        background-repeat: no-repeat;
        border: 1px solid #000000;
        height: 44px;
        overflow: hidden;
        width: 44px;
      }
    </style>
  </head>
  <body>
    <div></div>
  </body>
</html>
```

Designing your first icon

We're only going to be creating a single icon, so we won't quite need all of the empty space around the icon. Let's start by deciding what we want to illustrate. Our application is called Notekeeper and it creates notes. Perhaps an icon illustrating a sheet of paper would work?

This would have the added benefit of being fairly easy to represent at a small size. In the image editor of your choice, create a new document at 36 pixels by 36 pixels at 72 dpi. Name it as `notekeeper-icon-black-36.png`.

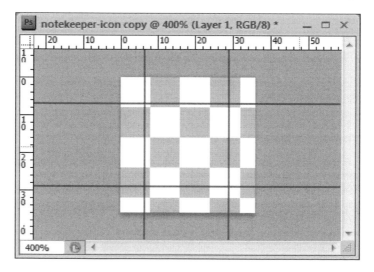

Even though the dimensions of the document are 36 by 36, the active area of the icon will only be 22 px by 22 px. This is in keeping with the icons provided by the jQuery Mobile team, and will make sure our icon doesn't look odd. To make it easier to stay within the lines, use the rectangular selection tool to draw a square at 22 px, then position it 7 px from the top edge of the document, and 7 px from the left. Next, draw guides along each edge so that your document looks something like this:

When drawing icons, you want to keep in mind the dimensions and attributes of the thing being illustrated. You won't be able to represent everything, but you need to communicate the spirit of the thing. A sheet of paper is taller than it is wide, and has lines on it. Let's start with those two things and see what we can come up with. The other icons in the set have a thick feel to them so that they can stand out against the background. Let's color in a solid shape, then delete the lines for the page so that the icon has the same thick feel. We're going to draw the lines in black so that they show up better printed in the book, but our icons will need to be white. Make sure you adjust your design accordingly.

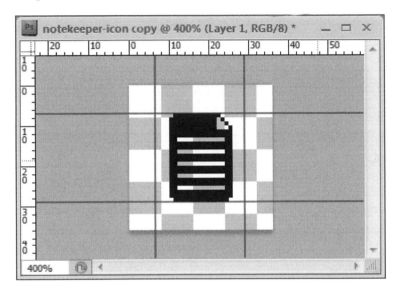

This icon seems to meet all of our criteria. It's taller than it is wide, and has lines just like paper. It also has a jaunty little page turn to give it some attitude. Isn't that what everyone looks for in their paper icon? Make sure that the icon's lines are white, then save it. The jQuery Mobile icons have been saved as transparent PNG-24 files. This is similar to the GIF format, but isn't required. Use transparent GIF if you wish.

When we created our first icon, we created the high resolution version. For brevity's sake, we're going to quickly walk through the steps of creating a low-resolution icon:

1. Create a new image document at 18 px by 18 px. Name this one `notekeeper-icon-18`.

2. The active area of this icon will be 12 px by 12 px. Draw a selection 12 px square, then position it 3 px from the top, and 3 px from the left.

3. Draw your guides, then sketch out the icon using the previous version as a reference. It's a lot harder drawing with this little space isn't it?

4. Your final result should look similar to the following screenshot:

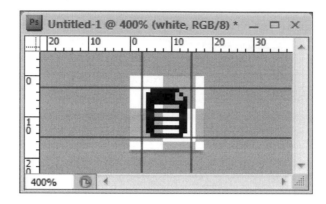

Save both images along with your Notekeeper theme then close Photoshop.

High and low resolution

Resolution is the number of dots or pixels that can be displayed into a given area. Those of you from the web world measure everything in 72 dpi, because that's what most monitors display. If you have much experience with mobile then you might know that each device can have a different resolution from those next to it. The problem with this is that higher resolution devices simply display more pixels on the screen. This means that an image displayed on a high resolution screen will be smaller than the same image on a low resolution screen.

jQuery Mobile accounts for this by having two versions of each icon, along with two sets of code for high and low resolution devices. In the next section we'll apply our custom theme and custom icon to our Notekeeper application.

Resolution independence

While it's nice that we can provide a better experience for users with higher resolution devices, we need to be concerned about devices that might have even higher resolutions. Those devices are out already, namely the Retina MacBook Pro and the iPad 3, both of which sport monstrous resolutions. Wouldn't it be nice if we could use a single file which scaled to whatever resolution the user's device had?

A great solution is using the CSS `@font-face` declaration: `http://caniuse.com/#search=@font-face`. By using an icon font you can centralize your icons into a single file, and you also get vector based crispness. This idea is so appealing that I took the time to build out a project that allows you to create your own icon font from a list of available icons. It's specifically made for jQuery Mobile, and spits out a few

CSS files, a few font files, and all of the code you need to integrate these icons into your jQuery Mobile project. Visit the following URL to try out the project yourself:

```
http://jqmiconpack.andymatthews.net/.
```

Updating the Notekeeper app

It's time for us to tie all of these loose ends together. We have a custom theme that we built using ThemeRoller, we've got our sweet custom icon, and now it's time for us to put all the pieces together. You'll need the following pieces to finish up:

- The code you completed at the end of the Notekeeper chapter
- The custom theme you created earlier in this chapter
- Your custom icon in white in both 18 px and 36 px sizes

Adding our custom theme

Let's start with the easiest part. Adding in our custom theme is pretty simple. Open notekeeper.html in your browser, and in the text editor of your choice. Additionally, open the index.html file from your theme download. We're going to merge the <head> tag from the theme file into notekeeper.html. You can see the results of the merge in the following code:

```
<title>Notekeeper</title>
<meta name="viewport" content="width=device-width, initial-
  scale=1">
<link rel="stylesheet" href="themes/Notekeeper.min.css" />
<link rel="stylesheet" href="styles.css" />
<link rel="stylesheet"
  href="http://code.jquery.com/mobile/1.3.1/jquery.mobile.structure-
  1.3.1.min.css" />
<script src="http://code.jquery.com/jquery-1.8.2.js"></script>
<script src="http://code.jquery.com/mobile/1.3.1/jquery.mobile-
  1.3.1.min.js"></script>
<script src="js/application.js"></script>
```

The first new line implements the new theme we created. The second line currently points to a missing file (because we haven't created it yet). Even with a rich theming system such as jQuery Mobile has, we're still going to have some custom CSS for various things. styles.css is where we'll put our assorted styles, especially the definitions for our custom icon.

By the way, go ahead and reload your browser window and take a look at our new theme in action. Isn't it snazzy? It's going to look even snazzier in a few minutes when our custom icon appears.

The astute among you might notice the deleted line. When ThemeRoller was first launched it provided a theme file that completely the entire default theme, even though the code contained in the default theme was being completely overridden. This meant that users would be downloading code they weren't ever going to use. When jQuery Mobile 1.1 was released, the team corrected that error by providing a `structure.css` file along with the theme.

Adding our custom icon

Go ahead and create `styles.css` in the root of your Notekeeper application code, and then open it. The first thing we'll do is to add in the declaration for our 18 px icon. It's low-resolution and will be the one you'll see in your desktop browser. High-resolution icons only show up in iPhone 4 and iPhone 4S at the moment.

To add our custom icon, we follow the pattern set by jQuery Mobile. It applies icons to buttons and other elements using the `.ui-icon` prefix. This means that for our icon to work within the framework, we have to name our CSS class as follows:

```
.ui-icon-notekeeper-note {
  background-image: url("themes/images/notekeeper-icon-white-
    18.png");
}
```

Then adding the icon to our **Add Note** button is as simple as adding a `data-icon` attribute like the following code:

```
<div class="ui-block-b">
  <input id="btnAddNote" type="button" value="Add Note" data-
    icon="notekeeper-note" />
</div>
```

Keep in mind that the string `notekeeper-note` can be anything as long as it matches the second half of the CSS class you created earlier. Finally, let's add the remaining piece to our app, the high-resolution icon.

One of the hallmarks of jQuery Mobile is its support for CSS media queries. Media queries allow you to query a given device for various pieces of information based on its media type: `screen`, `print`, `tv`, `handheld`, and several others. This answer to this query allows developers to branch CSS code and display the page one way for a desktop browser (`screen`), and another way for a TV (`tv`). In the case of our icons, we want to ask any viewing device with a type of screen if it supports a property called `-webkit-min-device-pixel-ratio` and if the value of that property is 2. Add the following lines to `styles.css` after the declaration for the low-res icon.

```
@media only screen and (-webkit-min-device-pixel-ratio: 2) {
  .ui-icon-notekeeper-note {
    background-image: url("themes/images/notekeeper-icon-white-
       36.png");
    background-size: 18px 18px;
  }
}
```

Other than the media query code, the only thing unique about this is the `background-size` property. It allows developers to specify that a given background should be scaled to the specified size (18 px by 18 px), rather than its original size of 36 px by 36 px. Since the resolution on the iPhone 4 and higher is exactly double the size of the low-resolution, this means that we're packing double the pixels into the same space as the smaller icon. The end result is that the icon looks crisper and sharper. If you've got one of these devices, upload your code to a server and view it. Your patience will be rewarded.

Summary

In this chapter we learned about advanced CSS techniques that are central to the jQuery Mobile experience, and how jQuery Mobile uses them to provide a rich interface to the end user. We took a deep dive into the basics of jQuery Mobile theming and how it works. We built a custom theme using the ThemeRoller tool, a custom icon with our very own hands, and we learned how to tie all those things together and implement them in our application

In the next chapter, you'll learn how to take the principles you've learned in the past 11 chapters and create a native application that can run on the iOS and Android platforms (along with several others), using the PhoneGap open source library.

12
Creating Native Applications

In this chapter, we will look at how to turn jQuery Mobile-based web applications into native applications for mobile devices. We'll discuss the PhoneGap framework, and how it allows you to tap into your device's hardware.

We will cover the following aspects:

- Discuss the PhoneGap project and what it does
- Demonstrate how to use PhoneGap's Build service to create native applications

HTML as a native application

For most folks, creating a native application on a platform like Android or iOS requires learning an entirely new programming language. While it is always good to learn new languages and expand your skill set, wouldn't it be cool if you could take your existing HTML skills, and use them natively on a mobile device?

Luckily, there is just such a platform. PhoneGap (http://www.phonegap.com) is an open-source project that allows you to take HTML pages and create native applications. This code is entirely free and can be used to develop applications for iOS (both iPhone and iPad), Android (again both phones and tablets), Blackberry, WebOS, Windows Phone 7, Windows Phone 8, Symbian, and Bada. PhoneGap works by creating a project in the native environment and pointing to an HTML file. Once set up, you can use your existing HTML, CSS, and JavaScript skills to create the UI and functionality of your application.

Even better, PhoneGap provides additional APIs to your JavaScript code. These APIs include the following:

- **Accelerometer**: It allows your code to detect basic movement on the device
- **Camera**: It allows your code to work with the camera

- **Compass**: It gives you access to the compass on the device
- **Connection**: It lets your application determine if your user is online, and if so, what type of connection is supported
- **Contacts**: Provides basic search and contact creation support
- **Device**: Basic device metadata like the operating system
- **Events**: Various types of events
- **File**: Read/write access to the device's storage
- **Geolocation**: Provides a way to detect the location of the device
- **Globalization**: Automatically formatting numbers and dates in your user's locale
- **Media**: Allows for basic video/audio capture support
- **Network**: Determines the network connectivity settings of the device
- **Notification**: A simple way to create a notification (via a pop up, sound, or vibration)
- **Storage**: Access to a simple SQL database

By using these APIs, you can take normal HTML sites and turn them into powerful, native-like applications that users can download and install on their devices.

Before we go any further, you should know that PhoneGap is actually the implementation of Apache Cordova. PhoneGap is Adobe's implementation, but for all intents and purposes, is the same thing. Since most people know the PhoneGap name, that is the one we will use in the book. Finally, do not forget that Cordova is free, open source, and available to all!

Working with PhoneGap

Creating a PhoneGap project is done via two main methods. The primary way people use PhoneGap is by using a command line tool and SDKs of the platform they are planning to support. The PhoneGap Docs (`http://docs.phonegap.com/en/3.0.0/guide_platforms_index.md.html#Platform%20Guides`) provides details on how to set up your environment for the device platform of your choice.

PhoneGap Documentation

Getting Started Guides

API Reference

Accelerometer

Camera

Capture

Compass

Connection

Contacts

Device

Events

Getting Started Guides

- Getting Started with Android
- Getting Started with BlackBerry
- Getting Started with iOS
- Getting Started with Symbian
- Getting Started with WebOS
- Getting Started with Windows Phone 7
- Getting Started with Windows Phone 8
- Getting Started with Windows 8
- Getting Started with Bada
- Getting Started with Tizen

Detailing the setup for each platform would be too much for this book (and would just duplicate what's on the PhoneGap website), so instead we will focus on the other option for creating native applications, the **PhoneGap Build** service. PhoneGap Build (`https://build.phonegap.com`) is an online service that simplifies and automates the process of creating native applications. It allows you to simply upload code (or use a public source control repository) to generate native binaries. Even better, you can use PhoneGap Build to generate binaries for all their supported platforms. That means you can write your code once and spit out code for an iPhone, Android, Blackberry, and other platforms, all from the site itself.

The PhoneGap Build service is not free, though. Pricing plans and other details may be found on the site, but luckily there is a free developer plan. That is the service we'll be using for this chapter. You can sign up with either an Adobe ID or GitHub credentials. After signing up you can then create your first PhoneGap Build project.

Notice that the Build service supports seeding a project from an existing Git or subversion repository or via an uploaded ZIP file. At this point, let's switch away from the website and back to code. We want to begin with a very simple set of code. Later on in the chapter we will do something a bit more interesting, but for now, our objective is to just upload some HTML and see what comes next. In the code you downloaded from GitHub, open the `c12` folder and look at the `app1` folder. This contains a copy of one of the list examples from *Chapter 4, Working with Lists*. It uses jQuery Mobile to create a simple list of four people, along with thumbnail pictures. Nothing too exciting, but it gets the job done for our purposes here. You will notice that there is already an `app1.zip` file.

If you go back to the website and click on **Upload an archive**, you can then browse to the location on your computer where you extracted the files and select that ZIP file. Be sure to also enter a name for the application; I chose `FirstBuildApp`. After clicking on **Ready to build**, you are then taken to a page with all your apps, which if you are a new Build user, will only contain the one just created:

Clicking on the app title then gives you the option to download various flavors of the application. Believe it or not, you are already able to download a version for most platforms. Working with iOS requires you to provide a certificate. If you create a certificate (which must be done on a Mac), you can upload it to PhoneGap Build and tell it to use the certificate for iOS builds. What's cool is that once you've done that, you can use a Windows (or even Linux) machine to work on your HTML, upload to PhoneGap Build, and create iOS builds. Click on the application title to go to the application's detail page:

Note that you can now download (again, you will need a certificate for iOS, or Blackberry if you choose) by simply clicking on the link next to your platform.

Actually using the applications depends on your platform of choice. For Android, you need to ensure that you have enabled the setting, **Allow installation of non-Market applications**. The exact wording and location of that setting will depend on your device. You can sign the application by editing the settings on the PhoneGap Build site. Once you've done that, you can actually submit your application to the Android Market. But since Android allows you to play with applications that are not signed, you can skip that step while testing. The following screenshot shows the application running on an Android:

And here is the same application running on an iPhone:

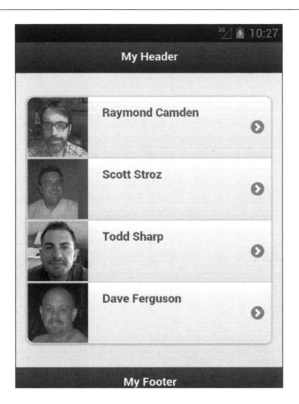

Adding PhoneGap functionality

We just demonstrated how to use the PhoneGap Build service to turn HTML (and JavaScript, CSS, and images of course) into a real, native application for multiple platforms. As mentioned earlier in the chapter, though, PhoneGap provides more than a simple wrapper to turn HTML into native applications. The PhoneGap JavaScript API provides access to a number of cool device-centric services that can greatly enhance the power of your application. For our second example, we'll take a look at one of these features—the **Contacts API**. (For full details, see the Contacts API documentation which is available at: `http://docs.phonegap.com/en/3.0.0/cordova_contacts_contacts.md.html#Contacts`).

The application in `Code 12-1` is a simple contact search tool. Let's take a look at the code and then cover what's going on:

```
Code 12-1: index.html
<!DOCTYPE html>
<html>
  <head>
    <title>Contact Search</title>
    <meta name="viewport" content="width=device-width, initial-
      scale=1">
    <link rel="stylesheet" href="jquery.mobile.min.css" />
    <script src="jquery.js"></script>
    <script src="jquery.mobile.min.js"></script>
    <script src="phonegap.js"></script>
    <script>
      document.addEventListener("deviceready", onDeviceReady,
        false);
      function onDeviceReady(){
        $("#searchButton").on("touchend", function() {
          var search = $.trim($("#search").val());
          if(search == "") return;
          var opt = new ContactFindOptions();
          opt.filter = search;
          opt.multiple = true;
          navigator.contacts.find(["displayName","emails"],
            foundContacts, errorContacts, opt);
        });
        foundContacts = function(matches){
          //create results in our list
          var s = "";
          for (var i = 0; i < matches.length; i++) {
            s += "<li>"+matches[i].displayName+"</li>";
          }
          $("#results").html(s);
          $("#results").listview("refresh");
        }
        errorContacts = function(err){
          navigator.notification.alert("Sorry, we had a problem and
gave
            up.", function() {});
        }
      }
    </script>
  </head>
  <body>
```

```
<div data-role="page">
  <div data-role="header">
    <h1>Contact Search</h1>
  </div>
    <div data-role="content">
      <input type="search" id="search" value=""  />
      <button id="searchButton">Search</button>
      <ul id="results" data-role="listview" data-
        inset="true"></ul>
    </div>
  </div>
</div>
</body>
</html>
```

Let's begin by looking at the layout portion of the application which resides in the bottom half of the file. You can see our jQuery Mobile page structure, and within it, an input field, a button, and an empty list. The idea here is that the user will enter a name to search for, hit the button, and the results will show up within the list. The following screenshot demonstrates the output:

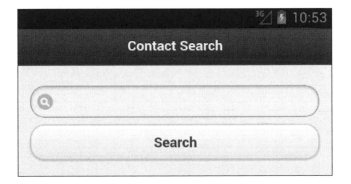

Now take a look at the JavaScript code. The first change we've made is to include the PhoneGap JavaScript library:

```
<script src="phonegap.js"></script>
```

You may be wondering, why is this file? You don't need it! When you upload your code to PhoneGap Build, the service automatically injects the proper JavaScript file for the platform.

The next interesting tidbit is the following line of code:

```
document.addEventListener("deviceready", onDeviceReady, false);
```

The `deviceready` event is a special event fired by PhoneGap. It essentially means that your code can now make use of advanced functionality, such as the Contacts API.

Within the event handler `onDeviceReady`, we have a few things going on. The first function of note is the event handler for the search button. The first few lines simply get, trim, and validate the value.

After we are sure there's actually something to search for, you can see the first actual use of the Contacts API, as shown in the following code snippet:

```
var opt = new ContactFindOptions();
opt.filter = search;
opt.multiple = true;
navigator.contacts.find(["displayName","emails"], foundContacts,
  errorContacts, opt);
```

The Contacts API has a search method. Its first argument is an array of fields to both search and return. In our case, we are saying we want to search against the name and e-mail values for contacts. The second and third arguments are the success and error callbacks. The final option is a set of options for the search. You can see it created before the call. The filter key is simply the search term. By default, contact searches return one result, so we specifically ask for multiple results as well.

Now let's take a look at the success handler:

```
foundContacts = function(matches) {
  //create results in our list
  var s = "";
  for (var i = 0; i < matches.length; i++) {
    s += "<li>"+matches[i].displayName+"</li>";
  }
  $("#results").html(s);
  $("#results").listview("refresh");
}
```

The result of the contact search will be an array of results. Remember that you only get back what you asked for, so our result objects contain the `displayName` and `emails` property. For now, our code simply takes the `displayName` and adds it to the list. Remembering what we learned from one of the previous chapters, we also know that we need to refresh the jQuery Mobile listview whenever we modify it. The following screenshot shows a sample search:

Summary

In this chapter we looked into the PhoneGap open-source project, and how it allows you to take your HTML, JavaScript, and CSS, and create native applications for a multitude of different devices. We played with the Build service and used it to upload our code and download compiled native applications. While jQuery Mobile isn't required with PhoneGap, the two make an incredibly powerful team.

In the next chapter, we'll take this team and create our final application, a full-fledged RSS reader.

13
Becoming an Expert – Building an RSS Reader Application

Now that you've been introduced to jQuery Mobile and its features, it's time to build our final, full application—an RSS Reader.

In this chapter, we will cover the following aspects:

- Discuss the RSS Reader application and its features
- Create the application
- Discuss what could be added to the application

RSS Reader – the application

RSS (Really Simple Syndication) is a way for sites to create a computer-readable index of their information. By using a common XML format, sites can let people read their content via other sites and applications. RSS is most popular on blog and news sites.

Before diving into the code, it may make sense to quickly demonstrate the application in its final working form, so you can see the pieces and how they work together. The RSS Reader application is exactly that; an application meant to take RSS feeds (for example from CNN, ESPN, and other sites), parse them into readable data, and provide a way for the user to view the articles. This application will allow you to add and delete feeds, providing both a name and a URL, and then provide a way to view the current entries from the feed.

The application begins with a basic set of instructions. These instructions are only visible when you run the application without any known feeds:

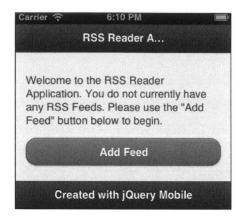

Clicking on the **Add Feed** button brings you to a simple form allowing for both a name and a URL (Unfortunately the URL has to be typed in manually. Luckily modern mobile devices allow for copy and paste. I'd strongly recommend using that!).

After adding the feed, you are returned back to the home page. The following screenshot shows the view after a few feeds are added:

To begin reading entries, the user simply selects one of the feeds. This will then fetch the feed and display the current entries, as shown in the following screenshot:

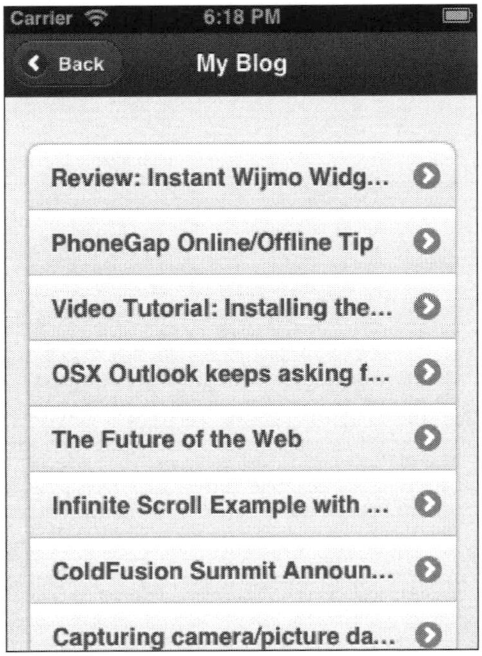

The final part of the application is the entry view itself. Some blogs don't provide a full copy of the entry via RSS, and obviously you may want to comment on the blog itself. So, at the bottom we provide a simple way to hit the real website, as shown in the following screenshot:

Now that you've seen the application, let's build it. Once again we're going to use PhoneGap Build to create the final result, but this application will actually run as is on a regular website as well.

Creating the RSS Reader application

Our application begins with the first page, `index.html`. This page will load in jQuery and jQuery Mobile as well. Its core mission is to list your current feeds, but it has to recognize when the user has no feeds at all and provide a bit of text encouraging them to add their first feed:

```
Code 13-1: index.html
<!DOCTYPE html>
<html>
  <head>
    <title>RSS Reader App</title>
    <meta name="viewport" content="width=device-width, initial-
      scale=1">
    <link rel="stylesheet" href="jquery.mobile/jquery.mobile-
      1.3.2.min.css" />
    <script src="jquery.mobile/jquery-1.9.1.min.js"></script>
    <script src="jquery.mobile/jquery.mobile-
      1.3.2.min.js"></script>
    <script src="main.js"></script>
  </head>
```

```
<body>
  <div data-role="page" id="intropage">
  <div data-role="header">
    <h1>RSS Reader Application</h1>
  </div>
  <div data-role="content" id="introContent">
    <p id="introContentNoFeeds" style="display:none">
      Welcome to the RSS Reader Application. You do not
        currently have any RSS Feeds. Please use the "Add Feed"
        button below to begin.
    </p>
    <ul id="feedList" data-role="listview" data-inset="true"
      data-
      split icon="dclctc"></ul>
    <a href="addfeed.html" data-role="button" data-theme="b">Add
      Feed</a>
  </div>
    <div data-role="footer">
      <h4>Created with jQuery Mobile</h4>
    </div>
  </div>
  <script>
      $(document).on("pagecreate", "#intropage", function(e) {
      init();
    });
  </script>
  </body>
</html>
```

As mentioned before in the code, we need to load up our jQuery and jQuery Mobile
templates first. You can see that in the beginning of the previous code. Most of the
remaining part of the page is boiler-plate HTML you saw in the previous chapter,
so let's call out a few specifics.

Firstly, make note of the introductory paragraph. Notice the CSS to hide the text?
The assumption here is that, most of the time, the user won't need this text,
as they will have feeds. Our code then is going to handle showing it when necessary.

Following that paragraph is an empty list that will display our feeds. Right below
that is the button that will be used for adding new feeds.

Finally, we've got a bit of script at the end. This creates an event listener for
the jQuery Mobile page event, pagecreate, that we tie into to then start up
our application tasks.

All of our code (our custom code that is) will be stored in `main.js`. This file is a bit big, so we'll simply show parts of it that relate to each section. Please keep that in mind as we go through the chapter. The entire file can be found with the rest of the book's sample code:

```
Code 13-2: Portion of main.js
function init() {
//handle getting and displaying the intro or feeds
  $(document).on("pageshow", "#intropage", function(e) {
    displayFeeds();
  });
```

Our first snippet from `main.js` comes from the `init` function. Remember, this is run on `pagecreate` for the home page. It's run before the page shows up. That makes it a good place to go ahead and register a function for when the page is displayed. We've taken most of that logic out into its own function, so let's take a look at that next.

The displayFeeds function

`displayFeeds` handles retrieving our feeds and displaying them. The logic is simple. If there are no feeds, then we want to display the introductory text. Otherwise we simply render out each feed:

```
Code 13-3: displayFeeds from main.js
function displayFeeds() {
  var feeds = getFeeds();
  if(feeds.length == 0) {
    //in case we had one form before...
    $("#feedList").html("");
    $("#introContentNoFeeds").show();
  } else {
    $("#introContentNoFeeds").hide();
    var s = "";
    for(var i=0; i<feeds.length; i++) {
      s+= "<li><a href='feed.html?id="+i+"' data-
        feed='"+i+"'>"+feeds[i].name+"</a> <a href=''
        class='deleteFeed'
        data-feedid='"+i+"'>Delete</a></li>";
    }
    $("#feedList").html(s);
      $("#feedList").listview("refresh");
  }
}
```

Notice we also clean out the list. It's possible a user had feeds and deleted them. By resetting the list to an empty string, we ensure that we don't leave anything behind. If there are feeds, we create the list dynamically, ensuring we call the `listview("refresh")` API at the end to ask jQuery Mobile to pretty up the list.

Storing our feeds

So where do the feeds come from? How do we store them? While we are using PhoneGap and could make use of the embedded SQLite database implementation, we can use something simpler — `localStorage`. `localStorage` is an HTML5 feature that allows you to store key/value pairs on the client. While you can't store complex data, you can use JSON serialization to encode complex data before it's stored. This makes storage of data extremely simple. Do keep in mind though, that `localStorage` involves file storage. Your application needs to read from a file whenever a change is made to the data. Since we are talking about a simple list of feeds, this data should be relatively small:

```
Code 13-3: getFeeds, addFeed, and removeFeed
function getFeeds() {
  if(localStorage["feeds"]) {
    return JSON.parse(localStorage["feeds"]);
  } else rcturn [];
}
function addFeed(name,url) {
  var feeds = getFeeds();
  feeds.push({name:name,url:url});
  localStorage["feeds"] = JSON.stringify(feeds);
}
function removeFeed(id) {
  var feeds = getFeeds();
  feeds.splice(id, 1);
  localStorage["feeds"] = JSON.stringify(feeds);
  displayFeeds();
}
```

The previous three functions represent the entire wrapper to our storage system. `getFeeds` simply checks `localStorage` for the value, and if it exists, handles converting the JSON data into a native JavaScript object using the `parse` function. `addFeed` takes a feed name and URL, creates a simple object out of it, and stores the JSON version. Finally, the `removeFeed` function simply handles finding the right item in the array, removing it, and storing it back to `localStorage`. Storage is done using the `stringify` function. As you can imagine, it takes data and turns it into a string.

Adding an RSS feed

So far so good. Now let's look at the logic necessary to add a feed. If you remember, the link we used to add a feed went to `addfeed.html`. Let's take a look at it:

```
Code 13-4: addfeed.html
<div data-role="page" id="addfeedpage" data-add-back-btn="true">
  <div data-role="header">
    <h1>Add Feed</h1>
  </div>
  <div data-role="content">
    <form id="addFeedForm">
    <div data-role="fieldcontain">
       <label for="feedname">Feed Name:</label>
       <input type="text" id="feedname" value=""  />
    </div>
    <div data-role="fieldcontain">
      <label for="feedurl">Feed URL:</label>
      <input type="text" id="feedurl" value=""  />
    </div>
    <input type="submit" value="Add Feed" data-theme="b">
  </div>
  <div data-role="footer">
    <h4>Created with jQuery Mobile</h4>
  </div>
</div>
```

There isn't much to this page outside of the form. Note that our form has no action. We aren't using a server here. Instead our code is going to handle picking up the form submission and doing something with it. Also note that we've not done something we recommended earlier — putting the jQuery and jQuery Mobile includes on top. Those includes are necessary in desktop applications because it's possible the user may bookmark a page outside of your application's home page. Since the eventual target for this code is a PhoneGap application, we don't have to worry about that. This makes our HTML files a bit smaller. Now let's return to `main.js` and look at the code that handles this logic.

The following code is a snippet from the `init` method of `main.js`. It handles the button click on the form:

```
Code 13-5: Add Feed event registration logic
//Listen for the addFeedPage so we can support adding feeds
$(document).on("pageshow", "#addfeedpage", function(e) {
```

```
    $("#addFeedForm").submit(function(e) {
      handleAddFeed();
      return false;
    });
  });
```

Now we can take a look at `handleAddFeed`. I've abstracted this code, just to make things simpler:

```
Code 13-6: handleAddFeed
function handleAddFeed() {
  var feedname = $.trim($("#feedname").val());
  var feedurl = $.trim($("#feedurl").val());
  //basic error handling
  var errors = "";
  if(feedname == "") errors += "Feed name is required.\n";
  if(feedurl == "") errors += "Feed url is required.\n";
  if(errors != "") {
    //Create a PhoneGap notification for the error
    navigator.notification.alert(errors, function() {});
  } else {
    addFeed(feedname, feedurl);
    $.mobile.changePage("index.html");
  }
}
```

For the most part, the logic here should be simple to understand. We get the feed name and URL values, ensure they aren't blank, and optionally alert any error. If an error didn't occur, then we run the `addFeed` method described earlier. Notice we make use of the `changePage` API to return the user to the home page.

I'll call out one particular bit of code here, the line that handles displaying the error:

```
navigator.notification.alert(errors, function() {});
```

This line comes from the PhoneGap API for notifications (`http://docs.phonegap. com/en/3.0.0/cordova_notification_notification.md.html#notification. alert`). It creates a mobile-specific alert notification for your device. You can think of it as a fancier JavaScript `alert()` call. The second argument is a callback function for the alert window dismissal. Because we don't need to do anything in that situation, we provide an empty callback that does nothing.

Viewing a feed

Moving on, what happens when a user clicks to view a feed? This is probably the most complex aspect of the application. We begin with the HTML template, which is rather simple because most of the work is going to be done in JavaScript:

```
Code 13-7: feed.html
<div data-role="page" id="feedpage" data-add-back-btn="true">
  <div data-role="header">
    <h1></h1>
  </div>
  <div data-role="content" id="feedcontents">
  </div>
  <div data-role="footer">
    <h4>Created with jQuery Mobile</h4>
  </div>
</div>
```

This page basically acts as a shell. Note it has no real content at all, just empty HTML elements waiting to be filled. Let's return to `main.js` and see how this works:

```
Code 13-8: Feed display handler (part 1)
//Listen for the Feed Page so we can displaying entries
$(document).on("pageshow", "#feedpage", function(e) {
  //get the feed id based on query string
  var query = $(this).data("url").split("=")[1];
  //remove ?id=
  query = query.replace("?id=","");
  //assume it's a valid ID, since this is a mobile app folks won't
    be messing with the urls, but keep
  //in mind normally this would be a concern
  var feeds = getFeeds();
  var thisFeed = feeds[query];
  $("h1",this).text(thisFeed.name);
  if(!feedCache[thisFeed.url]) {
    $("#feedcontents").html("<p>Fetching data...</p>");
    //now use Google Feeds API
    $.get("https://ajax.googleapis.com/ajax/services/feed/
      load?v=1.0&num=10&q="+encodeURI(thisFeed.url)+"&callback=?", {},
    function(res,code) {
    //see if the response was good...
    if(res.responseStatus == 200) {
```

```
        feedCache[thisFeed.url] = res.responseData.feed.entries;
        displayFeed( thisFeed.url);
      } else {
        var error = "<p>Sorry, but this feed could not be
          loaded:</p><p>"+res.responseDetails+"</p>";
        $("#feedcontents").html(error);
      }
    },"json");
  } else {
    displayFeed(thisFeed.url);
    }
  });
```

This first snippet handles listening for the pageshow event on feed.html. This means it will run every time the file is viewed, which is what we want since it is used for every different feed. How does that work? Remember that our list of feeds included an identifier for the feed itself.

```
for(var i=0; i<feeds.length; i++) {
  s+= "<li><a href='feed.html?id="+i+"' data-
    feed='"+i+"'>"+feeds[i].name+"</a> <a href=''
    class='deleteFeed'
    data-feedid='"+i+"'>Delete</a></li>";
}
```

jQuery Mobile provides us access to the URL via the data (url) API. Since this returns the entire URL and we only care about code after the question mark, we can use some string functions to clean it up. The end result is a numeric value query that we can use to fetch the data out of our feed query. In a regular desktop application, it would be pretty simple for a user to mess with the URL parameters. Therefore, we'd do some checking here to ensure that the value requested actually exists. Since this is a single user application on a mobile device, it really isn't necessary to worry about that.

Before we try to fetch the feed, we make use of a simple caching system. The very first line in main.js creates an empty object:

```
//used for caching
var feedCache= {};
```

This object will store the results from our feeds so that we don't have to constantly re-fetch them. That's why the following line is run before we do any additional network calls:

```
if(!feedCache[thisFeed.url]) {
```

So how do we actually get the feed? Google has a cool service called the Feed API (`https://developers.google.com/feed/`). It lets us use Google to handle fetching in the XML of an RSS feed and converting it to JSON. JavaScript can work with XML, but JSON is far easier since it becomes regular, simple JavaScript objects. We've got a bit of error handling, but if everything works well, we simply cache the result. The final bit is a call to `displayFeed`:

```
Code 13-9: displayFeed
function displayFeed(url) {
  var entries = feedCache[url];
  var s = "<ul data-role='listview' data-inset='true'
    id='entrylist'>";
  for(var i=0; i<entries.length; i++) {
    var entry = entries[i];
    s += "<li><a
      href='entry.html?entry="+i+"&url="+encodeURI(url)+"'>"+
      entry.title+"</a></li>";
  }
  s += "</ul>";
  $("#feedcontents").html(s);
  $("#entrylist").listview();
}
```

All that the previous block does is iterate over the result feed. When Google parsed the XML from the feed, it turned into an array of objects we can loop over. While there are a number of properties in the feed we may be interested in for the list, we care about the title only. Notice how we build our link. We pass the numeric index and the URL (which we will use in the next portion). This is then rendered to a simple jQuery Mobile listview.

Creating the entry view

Ready for the last part? Let's look at the individual entry display. As before, we'll begin with the template:

```
Code 13-10: entry.html
<div data-role="page" id="entrypage" data-add-back-btn="true">
  <div data-role="header">
    <h1></h1>
  </div>
  <div data-role="content">
    <div id="entrycontents"></div>
    <a href="" id="entrylink" data-role="button">Visit Entry</a>
  </div>
  <div data-role="footer">
```

```
        <h4>Created with jQuery Mobile</h4>
    </div>
</div>
```

Similar to `feed.html` before it, `entry.html` is an empty shell. Note that the header, the content, and the link are empty. All of these will be replaced with real code. Let's head back to `main.js` and look at the code that handles this page:

```
Code 13-11: Entry page event handler
$(document).on("pageshow", "#entrypage", function(e) {
    //get the entry id and url based on query string
    var query = $(this).data("url").split("?")[1];
    //remove ?
    query = query.replace("?","");
    //split by &
    var parts = query.split("&");
    var entryid = parts[0].split("=")[1];
    var url = parts[1].split("=")[1];
    var entry = feedCache[url][entryid];
    $("h1",this).text(entry.title);
    $("#entrycontents",this).html(entry.content);
    $("#entrylink",this).attr("href",entry.link);
});
```

So what's going on here? Remember that we passed an index value (which entry was clicked, the first or the second?) and the URL of the feed. We parse out those values from the URL. Once we know the URL of the feed, we can use our cache to get the specific entry. Once we have that, it's a simple matter to update the header, contents, and link. And that's it!

Going further

You can take the code from this application and upload it to the PhoneGap Build service now to try it out on your own device. But what else could we have done? Here's a short list of things to consider:

- PhoneGap provides a connection API (`http://docs.phonegap.com/en/3.0.0/cordova_connection_connection.md.html#Connection`) that returns information about the device's connection status. You could add support for this to prevent the user from trying to read a feed when the device isn't online.

- While we store the user's feeds in localStorage, the cached data from reading the RSS entry is stored temporarily. You could also store that data and use it when the user is offline.

- PhoneGap has an excellent plugin API and a great variety of plugins are already available. (http://plugins.cordova.io/) One of these plugins allows for easier sending of SMS messages. You could add an option to send an entry title and link to a friend via SMS. Did we mention PhoneGap also lets you work with your contacts? See the Contacts API for more information: http://docs.phonegap.com/en/3.0.0/cordova_contacts_contacts.md.html#Contacts.

Hopefully you get the idea. This is only one example of the power of jQuery Mobile and PhoneGap.

Summary

In this chapter, we took what we had learned about PhoneGap from the previous chapter and created a full, if rather simple, mobile application making use of jQuery Mobile for design and interactivity.

Index

E

echo.cfm template 66
entry view
  creating 214, 215
events
  about 147
  page events 153, 155
  physical events 147
events API 192

F

feed
  adding 210, 211
  storing 209
  viewing 212, 213, 214
feed API
  URL 214
fieldcontain wrapper 65
fieldset tag 68, 70, 74
file 192
  multiple pages, adding 20-22
five-column grids 84
fixed footers
  creating 35, 36
fixed headers
  creating 35, 36
flip toggle fields
  about 75
  creating 76
footers
  persisting 38, 39
  working with 33, 34
footer text 35
form
  about 63
  checkboxes, working 68-71
  flip toggle fields, creating 76
  jQuery Mobile, working with 64-68
  mini fields, working with 79, 80
  radio buttons, working 68-71
  search fields, using 75
  select menus, working 71-75
  slider fields, enabling 77, 78
form utilities, jQuery Mobile 142-145

four-column grids 84
full-screen footers
  creating 35, 36
full-screen headers
  creating 35, 36
functionalities
  adding to Notekeeper app, with JavaScript
    111-113

G

geolocation API 192
getFeeds function 209
globalization API 192
grids
  about 84
  content, laying out 84-88
  five-column 84
  four-column 84
  three-column 84
  two-column 84

H

headers
  adding 29-31
high resolution 186
home page 5-58
hotel mobile site
  building 55
  features 55
  home page 56-58
  hotel, contacting 61, 62
  hotel, finding 58, 59
  hotel rooms, listing 60
href attribute 123
HTML
  native application, creating from 191, 192
  writing 109-111
HTML page
  building 11-13

I

icon
  specifying 31
  using 47-49

load event
   pagebeforeload 153
   pageload 153
   pageloadfailed 153
loadNote() function 123
localStorage 114
   used, for storing Notekeeper data 114, 115
low-resolution icon
   about 186
   creating, steps 185

M

main.js file 208-215
matching swatches 169
media API 192
mini fields
   working with 79, 80
mixing swatches 169
mobile application
   about 105, 106
   reference link 105
mobile device
   resolution 186
mobileinit event 130
multiple files
   working with 23-25
multiple pages
   adding, to file 20-22

N

namespace 113
native application
   creating 191
   creating, from HTML 191, 192
native form controls
   using 78, 79
NavBars
   about 37
   footers, persisting across multiple pages 38, 39
   working with 37
navigation bars. *See* NavBars
network API 192
newPage variable 125

Notekeeper app 105
   add note wireframe, designing 108
   custom icon, adding 188, 189
   custom icon, designing 185, 186
   custom theme, adding 187, 188
   delete button wireframe 108
   designing 106
   functionalities, adding to 111-113
   HTML, writing 109-111
   notes wireframe, displaying 108
   requisites, listing 106, 107
   updating 187
   view note wireframe 108
   wireframe, building 107
Notekeeper data
   storing 114
   storing, localStorage used 114, 115
Notekeeper data, storing
   database 114
   localStorage 114
   pros and cons 114
   text file 114
note object 124
notes
   adding, to listview 121, 122
   deleting 126
   viewing 122
   viewing, on() method used 123
notesObj variable 118, 120
notes wireframe
   displaying 108
notification API 192

O

on() method
   used, for viewing notes 123
orientationchange event 148

P

pagebeforechange event 153
pagebeforecreate event 153, 157
pagebeforehide event 153
pagebeforeload event 153
pagebeforeshow event 153, 158
pagechange event 153

About Packt Publishing

Packt, pronounced 'packed', published its first book "*Mastering phpMyAdmin for Effective MySQL Management*" in April 2004 and subsequently continued to specialize in publishing highly focused books on specific technologies and solutions.

Our books and publications share the experiences of your fellow IT professionals in adapting and customizing today's systems, applications, and frameworks. Our solution based books give you the knowledge and power to customize the software and technologies you're using to get the job done. Packt books are more specific and less general than the IT books you have seen in the past. Our unique business model allows us to bring you more focused information, giving you more of what you need to know, and less of what you don't.

Packt is a modern, yet unique publishing company, which focuses on producing quality, cutting-edge books for communities of developers, administrators, and newbies alike. For more information, please visit our website: www.packtpub.com.

About Packt Open Source

In 2010, Packt launched two new brands, Packt Open Source and Packt Enterprise, in order to continue its focus on specialization. This book is part of the Packt Open Source brand, home to books published on software built around Open Source licences, and offering information to anybody from advanced developers to budding web designers. The Open Source brand also runs Packt's Open Source Royalty Scheme, by which Packt gives a royalty to each Open Source project about whose software a book is sold.

Writing for Packt

We welcome all inquiries from people who are interested in authoring. Book proposals should be sent to author@packtpub.com. If your book idea is still at an early stage and you would like to discuss it first before writing a formal book proposal, contact us; one of our commissioning editors will get in touch with you.

We're not just looking for published authors; if you have strong technical skills but no writing experience, our experienced editors can help you develop a writing career, or simply get some additional reward for your expertise.

Creating Mobile Apps with jQuery Mobile

ISBN: 978-1-78216-006-9   Paperback: 254 pages

Learn to make practical, unique, real-world sites that span a variety of industries and technologies with the world's most popular mobile development library

1. Write less, do more: learn to apply the jQuery motto to quickly craft creative sites that work on any smartphone and even not-so-smart phones.

2. Learn to leverage HTML5 audio and video, geolocation, Twitter, Flickr, blogs, Reddit, Google maps, content management system, and much more.

3. All examples are either in use in the real world or were used as examples to win business across several industries.

jQuery for Designers: Beginner's Guide

ISBN: 978-1-84951-670-9   Paperback: 332 pages

An approachable introduction to web design in jQuery for non-programmers

1. Enhance the user experience of your site by adding useful jQuery features.

2. Learn the basics of adding impressive jQuery effects and animations even if you've never written a line of JavaScript.

3. Easy step-by-step approach shows you everything you need to know to get started improving your website with jQuery.

Please check **www.PacktPub.com** for information on our titles

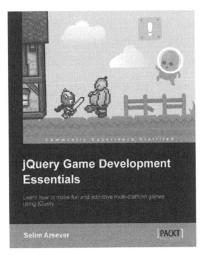

jQuery Game Development Essentials

ISBN: 978-1-84969-506-0   Paperback: 244 pages

Learn how to make fun and addictive multi-platform games using jQuery

1. Discover how you can create a fantastic RPG, arcade game, or platformer using jQuery!

2. Learn how you can integrate your game with various social networks, creating multiplayer experiences and also ensuring compatibility with mobile devices.

3. Create your very own framework, harnessing the very best design patterns and proven techniques along the way.

jQuery Hotshot

ISBN: 978-1-84951-910-6   Paperback: 296 pages

Ten practical projects that exercise your skill, build your confidence, and help you master jQuery

1. See how many of jQuery's methods and properties are used in real situations. Covers jQuery 1.9.

2. Learn to build jQuery from source files, write jQuery plugins, and use jQuery UI and jQuery Mobile.

3. Familiarise yourself with the latest related technologies like HTML5, CSS3, and frameworks like Knockout.js.

Please check **www.PacktPub.com** for information on our titles

Made in the USA
San Bernardino, CA
05 March 2014